Amal Toumi
Mounir Frija

Optimization and Simulation of 3D Printing of Medical Implants

Amal Toumi
Mounir Frija

Optimization and Simulation of 3D Printing of Medical Implants

Simulation of metal printing for hip and knee prostheses

ScienciaScripts

Cover image: www.ingimage.com

This book is a translation from the original published under ISBN 978-620-6-72630-2.

Publisher:
Sciencia Scripts
is a trademark of
Dodo Books Indian Ocean Ltd. and OmniScriptum S.R.L publishing group

120 High Road, East Finchley, London, N2 9ED, United Kingdom
Str. Armeneasca 28/1, office 1, Chisinau MD-2012, Republic of Moldova, Europe
Printed at: see last page
ISBN: 978-620-8-26554-0

Table of contents

Chapter 1: BIBLIOGRAPHICAL STUDY 4

Chapter 2: DIGITAL SIMULATION METHODOLOGY FOR IMPRESSING AN HIP PROSTHESIS 12

Chapter 3: Parametric optimisation of the SLM process 31

Chapter 4: TOPOLOGICAL OPTIMIZATION OF A KNEE PROSTHESIS FOR ADDITIVE MANUFACTURING 42

Conclusion and outlook 61

References 63

SUMMARY

In this book, we take you into the vast and innovative world of additive manufacturing and its applications, with a focus on powder-based laser fusion manufacturing of medical implants. Numerical simulation of the SLM/LBM process on a Titan Ti-6Al-4V superalloy femoral stem using **Simufact Additive** software is explored and the application of topological optimisation using lattice structures on a femoral component of a knee prosthesis using **nTopology** software is studied and validated by static analysis using **ANSYS** software.

CHAPTER 1: LITERATURE REVIEW

CHAPTER 1: BIBLIOGRAPHICAL STUDY

1 Introduction

1.1. Background to additive manufacturing

- **3D printing: The first years**

3D printing, technically known as additive manufacturing, was introduced in the mid-1980s by Charles Hull [1]an American engineer. It is simply a process for obtaining functional parts. However, unlike traditional subtractive manufacturing processes, where material is removed from a block to obtain the desired shape (machining, milling, etc.), 3D printing creates the object by depositing the material successively layer by layer, based on a digital model. Hull was able to make a small plastic cup by polymerising successive layers of liquid resin using an ultraviolet laser [2]. This was the first layer in the history of additive manufacturing. It has revolutionised the field of industrial production by enabling greater design flexibility, waste reduction and product customisation.

- **Necessity is the mother of invention**

Innovative manufacturing technologies are essential to meet the challenges faced by industry, such as reducing production costs, the need to design complex parts that are both lightweight and high-performance, and the growing demand for customised products. Additive manufacturing has proved to be the answer. It offers a particularly suitable solution. With this technique, shape and geometry are no longer an obstacle. FA has emerged in response to specific needs in various sectors, such as the automotive, aerospace and medical industries. It has opened the door to new designs, for example bio-inspired parts such as lattice structures to ensure the osseointegration of implants with the human body. So additive manufacturing represents not just a technological revolution, but also a pragmatic response to contemporary demands, proving that necessity stimulates innovation. However, it lacks scientific hindsight and a good grasp of these parameters on the part of users.

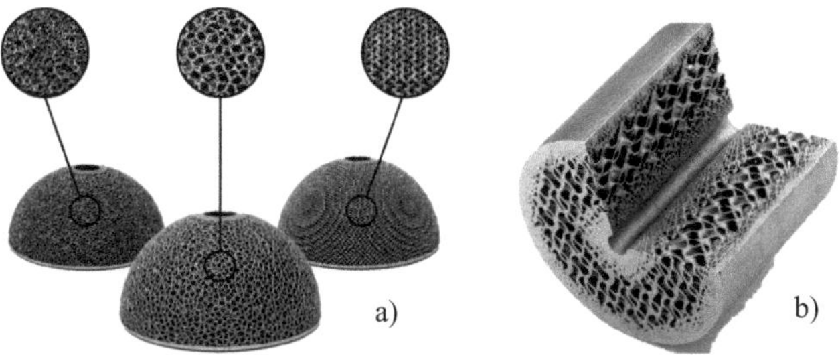

Figure 1 :a) Orthopaedic implant in different lattice structures using nTopology software ; b) Complex geometry not achievable using conventional manufacturing technologies.

1.2. SLM metal additive manufacturing

Several FA technologies can be used to produce metal parts, but additive manufacturing by laser fusion is currently the most widely used in industry. There are two types of laser melting FA process: the first is "Laser melting/sintering by powder projection", which involves melting the surface of a metal part using a laser while simultaneously projecting powdered material into the area, and the second is "Melting/sintering on a powder bed", which involves melting the material, which is deposited layer by layer to form a powder bed. This technique has various names in the literature: **laser beam melting (LBM)** or **Selective Laser Melting/Sinetering (SLM/SLS), laser powder bed fusion (LPBF), etc.** It is particularly well suited to high-end applications. Two sectors already seem to be making the most of it: biomedical and automotive, while a third field of application is emerging: flight parts for the aerospace industry.

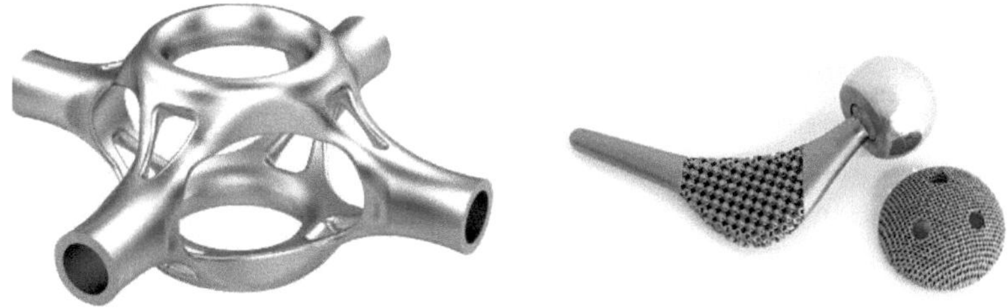

Figure 2: Parts manufactured by SLM: a) Motor gimbal b) Hip prosthesis

- **Principle and general methodology of the SLM process**

Selective laser melting (SLM/LBM) is a simple process for manufacturing metal parts. As with any manufacturing technique, this process involves a chain of operations. These can be summarised in the five sequences shown in **Fig.1.**

1. Designing a 3D digital model of the product to be manufactured using CAD software and converting it as an STL file.
2. The manufacturing process begins by melting the metal particles with the laser source before the roller applies a new layer of powder.
3. Successive layers of material are deposited until the required thickness is achieved and the object is fully printed.
4. Heat treatment is required to mitigate residual stresses in the metal part.
5. Separation of the cooled part from the tray and removal of supports and unfused powder particles.

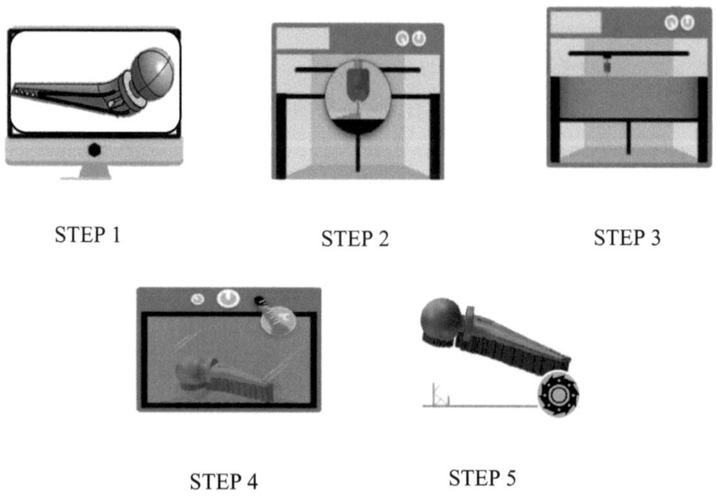

Figure 3: SLM manufacturing stage

SLM machines generally use fibre or diode lasers. The metal powder particles, which are between 15 and 45 µm in size, are placed in a controlled enclosure (inert atmosphere) to prevent oxidation: the gas is generally argon/nitrogen. [3]. **Figures 4** and **5** show a 3d laser fusion printing machine (CETIME).

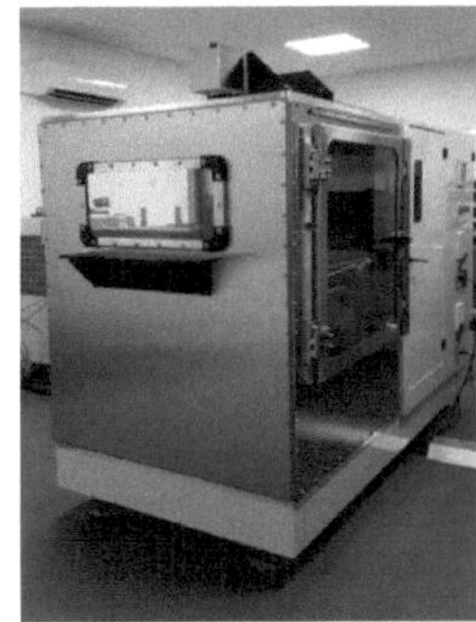

Figure 4: Powder bed additive manufacturing machine

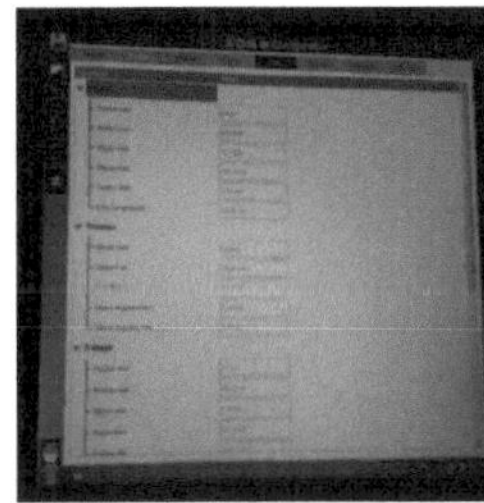

Figure 5: Configuration table for laser additive manufacturing machine

- **Materials**

SLM metal additive manufacturing allows the use of several materials, each with specific properties adapted to different applications. **Table 1 shows** a selection of the most commonly used materials. Titanium is widely chosen for medical implants because of several key properties and advantages. In addition to its light weight, titanium is known for its biocompatibility and corrosion resistance, particularly in biological environments.

Table 1: Different materials used by SLM

Materials	Properties	Domains
Titanium	• Resistant • Corrosion-resistant • Light • Biocompatible • Good adhesion to bone	Medical (prostheses, etc.) Aerospace (complex and/or multifunctional parts)
Stainless steel	• Good mechanical properties • Highly resistant to corrosion	Aerospace Rapid prototyping Stainless steel tools
Cobalt-chrome	• Good mechanical properties • Wear-resistant • Heat resistant	Medical sector (dental prostheses or crowns) Turbines
Aluminium	• Resistant • Light • Good thermal conductivity • Low density	Aerospace industry Engine
Nickel	• Mechanical resistance at high temperatures • Resistant to hot oxidation	Engine parts
Copper alloys	• Good electrical and thermal conductivity	Heat exchangers

2 Limits

However, as technology is still in its infancy, there are many constraints and limits to this freedom of design. The cost of manufacturing remains the most limiting factor, mainly due to the cost of machines and their low productivity. The need for supports can lead to costly post-processing. The quality of the final product depends on the experience of the parts manufacturer, the powder, the parameters and the machine used, etc. It is therefore important to control the entire process. Many of these limitations are gradually being reduced and we are seeing regular improvements in terms of productivity, machine size and the range of materials available, but at present these limitations are strong and can generate significant additional costs.

3 Digital simulation: a must

Numerical simulation of the SLM process has become crucial for overcoming process-related challenges. Technical issues such as control of mechanical parameters and properties, biocompatibility and surface quality can be anticipated and optimised using accurate numerical models. Simulation also provides rapid, cost-effective information on a range of aspects:

- **Construction layout:** print orientation, required support structures...
- **Deformation of parts:** the deformations and distortions that occur in a part as a result of thermomechanical phenomena...
- **Metrics relevant to the process:** Construction time, estimated use of materials, expected quantity of support material, etc.
- **Post-treatment planning:** The build-up of stresses and the thermal post-treatment that may be required to relax them.
- **Contingency planning:** Errors that will lead to complete failure (delamination of the part from the construction substrate, the appearance of defects or the formation of porosities).
- **Parameter optimisation:** finding the best compromise between parameters and part quality, while minimising manufacturing errors.

Depending on the objective in mind, there are essentially three scales of simulation.

- **Microscopic scale:** the model simulates laser beam-powder particle interaction
- **Mesoscopic scale:** Modelling is carried out at the scale of material deposition, in which melting on a powder bed, hydrodynamics in the melt and thermal stress are simulated.
- **Macroscopic scale:** This thermomechanical model deals with heat transfer and mechanical response (residual stresses and distortion) at the scale of the part (fluid flow in the melting zone is ignored).

4 Conclusion

This chapter has presented a general state of the art on additive manufacturing processes, with more detail on LBM manufacturing. We have also looked at terms related to the implantation of medical devices, such as biocompatibility, biomaterials, and so on. The materials that best meet these biocompatibility constraints were introduced, and we chose to work with a Ti6Al4V titanium alloy as our reference material because of its excellent mechanical properties, biocompatibility and good corrosion resistance.

CHAPTER 2: NUMERICAL SIMULATION METHOD FOR PRINTING A HIP PROSTHESIS

CHAPTER 2: DIGITAL SIMULATION METHODOLOGY FOR IMPRESSING AN HIP PROSTHESIS

1 Simulation methodology

In this book, the macroscopic scale is the one studied. Essentially, the areas where residual stresses are concentrated and the distortions existing in the part at the end of the manufacturing process will be dealt with. Firstly, the three-dimensional digital model of the process produced using Solidworks® software will be described.

Next, the model will be simulated using Simufact Additive® software and a study of the machine parameters (laser power, scanning speed, layer thickness) will be carried out.

1.1 Geometric model

The geometric model we are going to study, produced using SolidWorks software and imported from GRABCAD, consists mainly of three elements: a cup (acetabulum), an insert, and a femoral stem **fig.6**.

Component	Description	CAD
Cupule	A hemispherical acetabular component embedded in the acetabulum.	
Insert	Insert fixed in the cup, forming the friction couple with the ball on the stem.	

Femoral stem	Femoral component, consisting of a stem ending in a ball (monobloc), fixed in the femur bone.	

Only the femoral stem of the prosthesis will be simulated by the Simufact Additive software.

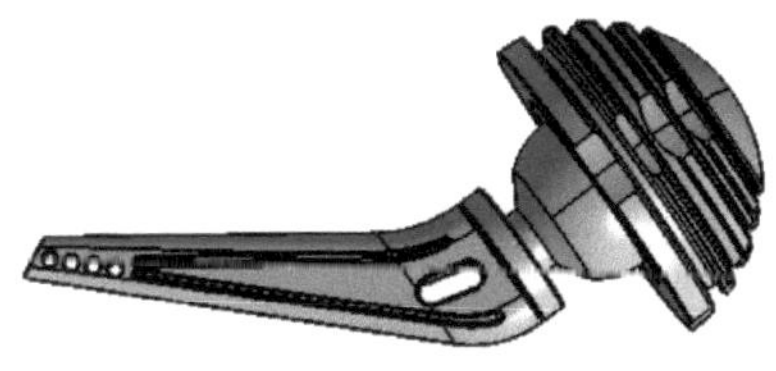

Figure 6 Geometric model of the PTH

1.2 Materials

Titanium and its alloys are popular metals for additive manufacturing because of their high strength, high mechanical resistance, low density, excellent corrosion resistance, high biocompatibility and good formability (ability to undergo deformation without damage). [4] Ti-6Al-4V is one of the weakly stabilised α-β alloys, its chemical composition is given **Table2.** The heat treatment response of this alloy is directly related to the amount of β phase after quenching and its instability, it exhibits good properties at room temperature due to its relatively high aluminium content. The alloy has different microstructures depending on the heat or thermomechanical treatment path. This gives Ti-6Al-4V a wide variety of properties, particularly mechanical. [5]

1.2.1 Chemical composition

Table 2: Chemical composition of Titanium Ti6Al4V

element	Composition % mass
Aluminium	6
Vanadium	4
Carbon	<0,08
Iron	<0,3
Oxygen	<0,2
Nitrogen	<0,07
Titanium	Base

1.2.2 Properties

Table 3: Titanium alloy properties [6]

Physical and thermal properties	Density in g/cm^3	4,3
	Melting temperature in °C	1648
	Transformation point: - Transus Beta in °C	1000 °C
	Thermal conductivity in W/m.°C at 20 °C	6,7
Mechanical properties	Modulus of elasticity in GPA at 20 °C	110
	Compression modulus GPa	115 - 121
	Shear limit GPA	42
	Poisson's ratio	0.33
	Torsional modulus in N/mm^2	45000
Electrical property	Electrical resistivity in μΩ.cm at 20 °C	170

1.3 Process parameters and phenomena

To obtain a healthy, high-performance part, you need to know, set and control some of the main parameters that can have the greatest influence on the manufacturing process. These parameters can be classified into five families illustrated in the diagram in **Fig9.** [7]

The laser-material interaction of the LBM process gives rise to various types of physical phenomena that are closely interlinked **figure.7** [7] [8]. The most dominant phenomena are illustrated in **figure8.**

- Thermal (convection, conduction, radiation, melting and solidification) ;
- Metallurgy (phase transformation, evaporation, changes in chemical composition, composition of materials, crystallographic structure) ;
- Mechanics (deformation, stress, laws of behaviour).

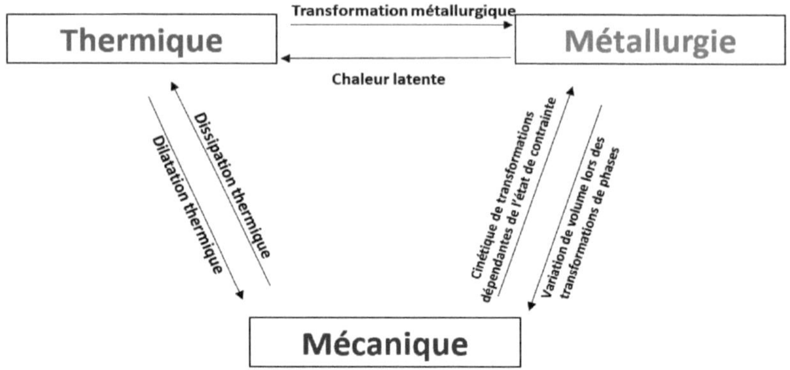

Figure 7 Coupling of physical phenomena

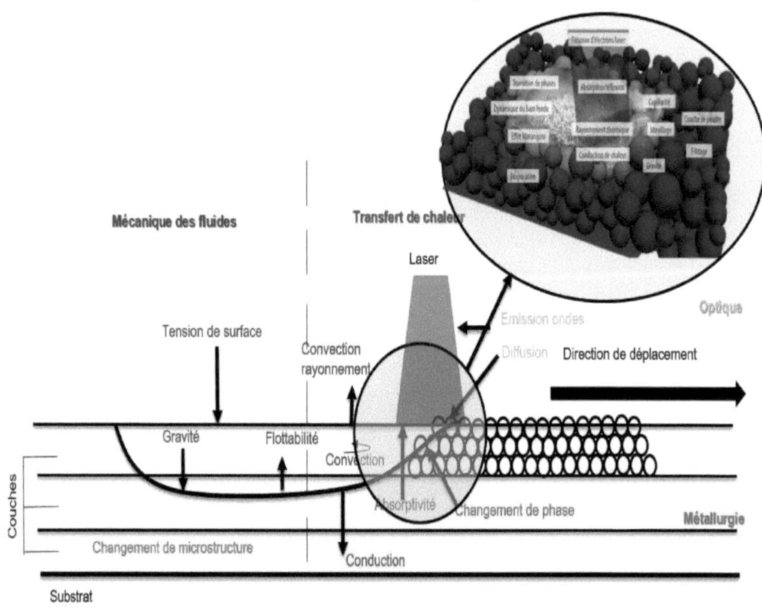

Figure 8 Summary of the physical phenomena involved

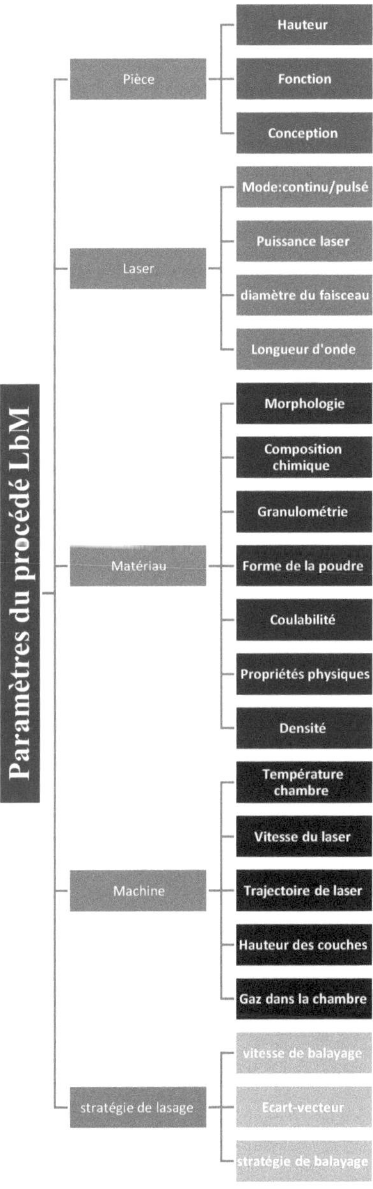

Figure 9: LBM process parameters [9]

2 Case study : Digital simulation of hip prosthesis

2.1 Choice of simulation software

Simulation programmes for the laser melting on powder bed process have recently been developed based on the latest research into additive manufacturing technology using this process. They are used to simulate heat transfer during the process and the propagation of heat between the metal layers, this type of simulation is used to better understand the heat cycles undergone by the part being manufactured; The aim here is to understand and control the deformation of the parts during LBM manufacturing and the residual stresses caused by the thermomechanical history undergone during the manufacturing process. Examples of programs used for this type of simulation are "Abaqus", "Simulia", "Amphyon", "Simufact Additive", "Simulia", "Amphyon", "Amphyon" and "Simulia". Simufact Additive .....

In our case, the simulation will be carried out by the Simufact Additive software, which specialises in the simulation and optimisation of SLS and SLM (LBM) additive manufacturing processes, fusion / sintering on a powder bed, taking into account the complete chain of key operations in the technology: from the creation (or import) of the supports, to the printing of the 3D part, via the stress-relieving heat treatment (post-printing), the cutting of the part from the platen, the cutting of the supports, and the treatment of the porosity by hot isostatic compression (CIC/HIP). It will mainly enable the prediction of residual part deformations and stresses during the additive manufacturing process. [10]

2.2 The different stages of simulation

- **Importing the part**

As with any 3D printing technique, the simulation of a part begins with the import of the 3D model (.STL format), designed using CAD software, into the workspace of the simulation software. The machine selected is RENISHAW-AM 250, with a build space of dimension (250, 250,300) **[APPENDIX 1].**

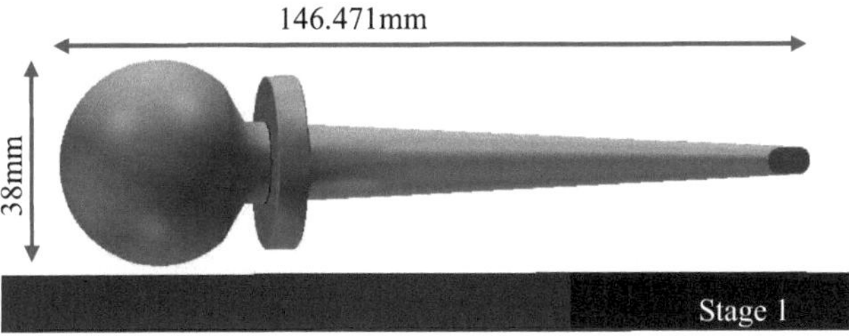

- **Media generation**

Manufacturing supports for 3D printing are an essential part of the design process. The supports make it possible to build surfaces of the part oriented above a certain angle, known as the cantilever angle, in relation to the manufacturing plate or the surfaces in the void. The degrees of these overhangs will depend on the ability of each 3D printer to support the overhangs. In most cases, printing overhangs of up to 45° is possible. Beyond that, supports are required.

These supports are also used to :

- Temporary support for the part under construction, which will be rigid but fragile during the process
- Heat dissipation ;
- Support newly merged surfaces,
- Prevention of part deformation due to residual stresses ;

The volume of support generated is equal to 16595.3 mm^3 , which is almost equal to 33% of the volume of the part which is 50581.8 mm .[3]

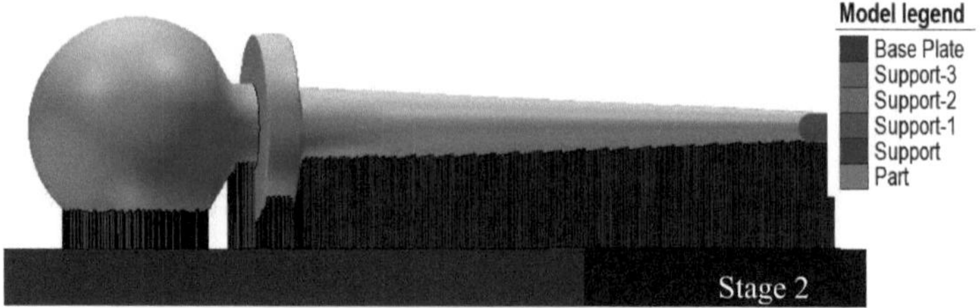

- **Definition of Ti-6Al-4V material**

The properties of titanium alloy powder are shown in the table below.

Table 4 Properties of titanium used in Simufact Additive

Density	4.41 kg/mm3
Melting temperature	1600 C°.
Solid temperature	1550 C°.
Latent heat	4.19 E^{11} mm^2/s^2
Particle size	15-45μm
Layer thickness	30 μm

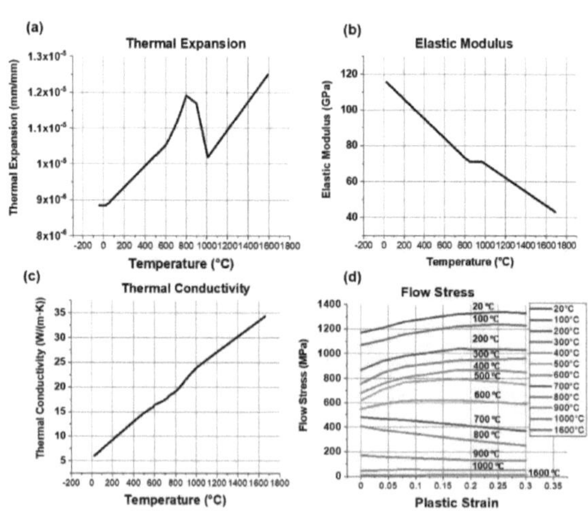

Figure 10 Thermo-physical-mechanical property curves for Ti-6Al-4V

- **Definition of machine stages and parameters**

The process chain in simufact is made up of different stages and can be customised to suit the user.

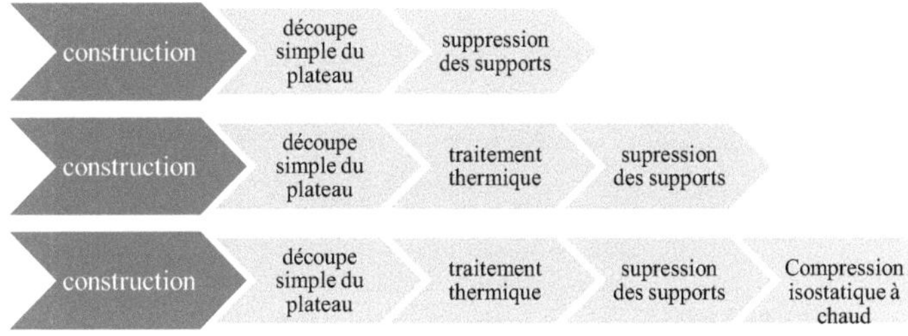

Figure 11 Simufact additive process chain

The aim here is, firstly, to know which areas are most affected by residual stress and distortion, and at what stage the stress reaches its maximum value. So it's sufficient to work with chain 1, which consists of depositing the layers, separating the part from the tray and then removing the supports.

Table 5 Machine parameters

Coat thickness (µm)	30
Laser power (W)	100
Sweep speed (m/s)	1
Laser beam diameter (µm)	50

- **Generation of the part mesh**

Mesh generation can be defined as the division of a physical body into small elements. In additive manufacturing we use voxels (volume+pixels).

In the vocabulary, voxel represents a value on a regular grid in a three-dimensional space. They are small cubes, on which the number of layers deposited for the construction of the part will depend. Some research into the effect of voxels on results has shown that as the size of the voxel element increases, the amount of distortion calculated decreases. The cubes in this case are **1mm** in size. The results of the number of voxels generated for each element are shown in the table.

Figure 12Voxel in Simufact additive

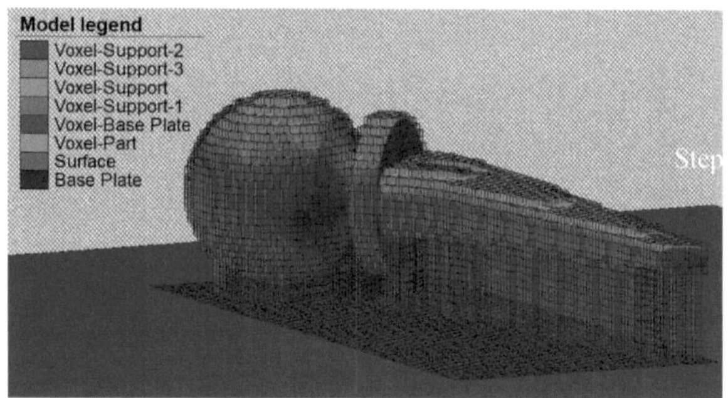

Figure 13 Part meshed by voxel

Table 6Number of voxels for each element

voxels in the room	17120
Voxel of the construction tray	3920
support voxels	7832
support voxels 1	308
support 2 voxels	215
support voxels 3	850

the number of layers produced is equal to 28 layers.

- **Launching the calculation and interpreting the results**

Once all the steps have been completed, all that's left to do is run the simulation and then move on to the phase of analysing and interpreting the results.

3 Simulation results

3.1 simulation without post-processing

The results obtained for the analysis of the distribution of stresses and distortion show that the maximum values are reached when machining the part from the building board. This can be explained by the sudden slowing down of stresses when the part is detached from the board. The part, which was previously under tension, is mechanically realigned, which can cause deformation and an increase in the measured stresses. The maximum stress value is 1330.36 MPa and the maximum distortion value is 5.34 mm.

Looking at the figures, the stresses induced on the part at the end of construction are highest at the level of the supports fixed to the plate. After removal of the supports and the plate, the stresses are intensively localised in the femoral head. This phenomenon is due to the effects of the compressive stresses generated by the head (maximum weight), which counteract the residual tensile stresses generated during additive manufacturing to create balance in the part.

The deformation of the manufactured part, generated by residual stresses and machining (cutting of the plate and removal of supports), compared with the initial part is clearly shown in Figure 14. This deformation has an impact on the mechanical strength of the part. They are more intense at the head than in the lower part of the stem. This can be justified by the fact that the femoral head generates the maximum residual stresses.

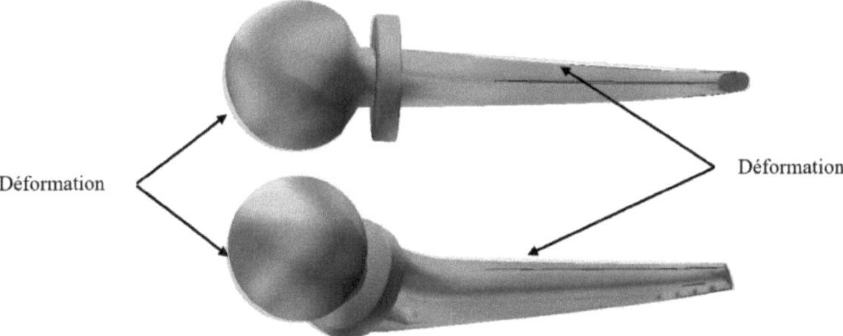

Figure 14Deformation of the manufactured part in relation to the final part

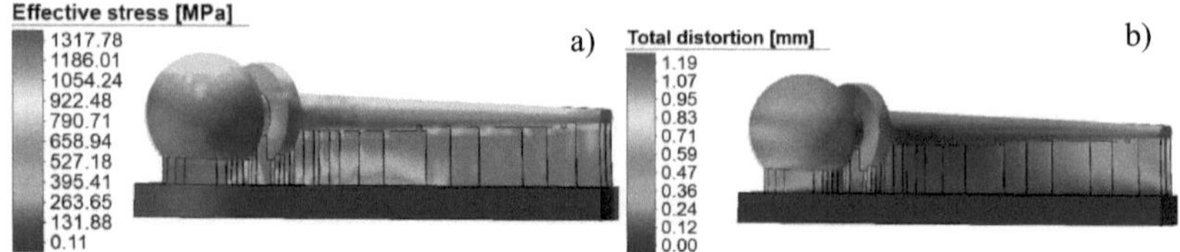

Figure 15State of stress (a) and distortion (b) in the part at the end of construction.

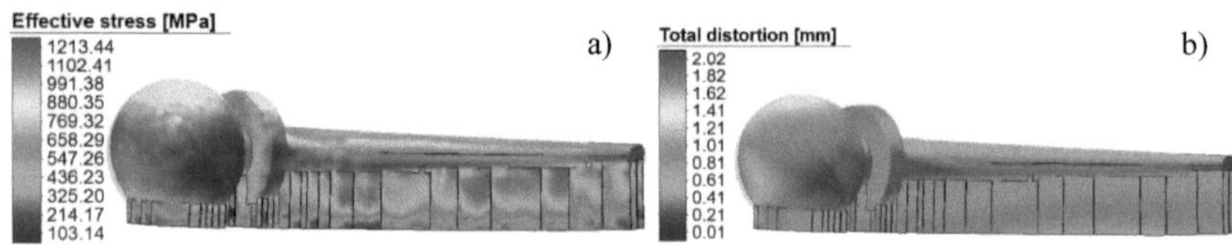

Figure 16 State of stress (a) and distortion (b) in the part after cutting the plate

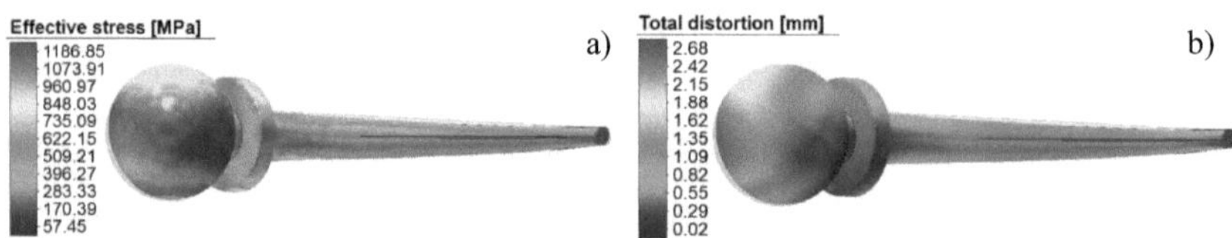

Figure 17 State of stress (a) and distortion (b) in the part after removal of the supports

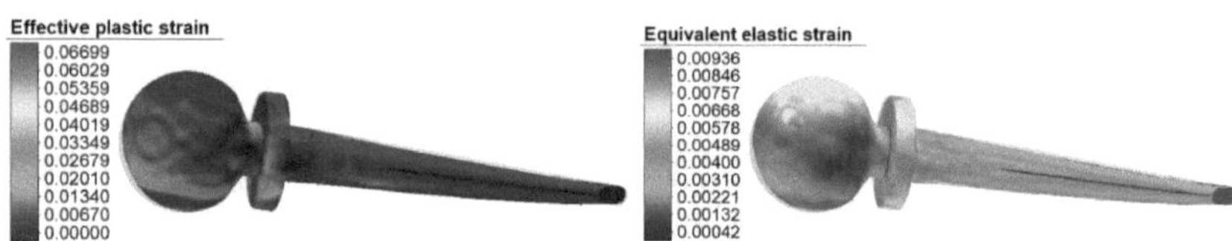

Figure 18 Characteristics of the final part

Table 7: Assessment table for induced effects based on manufacturing stages

	After construction	After machining	After removal of supports
Max. effective stress (MPa)	1317,78	1330,36	1186,85
Max. distortion (mm)	1,19	2,08	2,68
Max. shear stress (MPa)	625,34	603,83	581,18
Shear stress min (MPa)	-623,17	-669,69	-547,32

We need to understand the residual stresses generated during laser melting of metal powders in order to find ways of reducing or minimising them. These are internal stresses that exist, in a self-balancing way, in a free body (without external forces and stresses). They can be classified according to the mechanism that generates them. [11]

However, these residual stresses and strains are very complex to predict because of very high local temperatures, rapid heating and cooling cycles, and the mobility of the laser heat source.

During the SLM process, the energy supplied by the laser beam (100W), which advances at a speed of around 1 m/s, is absorbed by the powder bed and converted into heat, which cools very rapidly (cycles of expansion of the upper molten layers and contraction of the lower cold layers). This phenomenon produces very large temperature gradients in the manufactured part, and consequently the generation of residual stresses. There are 2 types of stress generated during laser passage: compressive stress in the heat-affected zone creating expansion, and tensile stress in the surrounding area. Then, during cooling, the phenomenon is reversed. The contraction of the cooled zone causes tensile stresses within it and compressive stresses around it.

The phase transformations undergone by the part during the process, such as allotropic phase transformations (changes in crystalline structure), also induce residual stresses, which can lead to distortion of the part, reduced mechanical strength and even the appearance of cracks.

3.2 Simulation with post-processing

The residual stresses generated in the continuous transient process of melting and solidification are still a major obstacle to the additive manufacturing of large, high-performance metal parts.

As a result, post-treatment to relieve stresses during or after the part construction process is necessary. Four types of heat treatment are listed in the literature and have been tested on Ti-6Al-4V titanium made by additive manufacturing: [12]

- ✓ Stress relaxation treatment or annealing
- ✓ Subtransus solution treatment
- ✓ Supersolvus solution treatment
- ✓ Hot Isostatic Compression treatment (HIP / CIC)

Simufact additive software offers two types of post-processing:

- Heat treatment by annealing (700°C for 3 hours) which improves certain mechanical properties (elongation, fatigue strength, reduces hardness, increases ductility) and facilitates the elimination of internal residual stresses.
- Hot isostatic pressing (CIC/HIP), which involves combining the effect of high pressure applied by a neutral gas (often argon at pressures in excess of 1,000 bar) with high temperature. This reduces or even closes internal porosities and improves certain properties of the material, such as tribological properties, fatigue resistance, corrosion resistance, etc.

Carrying out one or more of these heat treatments is sometimes essential to ensure good geometry and good mechanical properties in the manufactured part. It therefore seems necessary to study their effects on stresses and distortions.

3.3 The effect of treatment on constraints and distortions

The treatment conditions (cycle time, atmosphere conditions, furnace performance) depend on the material being treated, the objectives sought and the application. They vary enormously, from air treatment for an aluminium alloy to vacuum treatment for a titanium alloy.

This study simulates the effect of annealing treatment and CIC treatment on a part manufactured by laser fusion.

The following curves are associated with the heat treatment of the Ti6Al4V alloy:

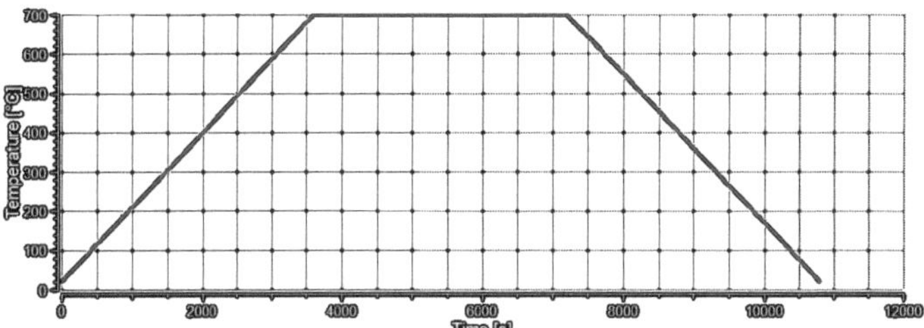

Figure 19 Temperature variation curve for heat treatment

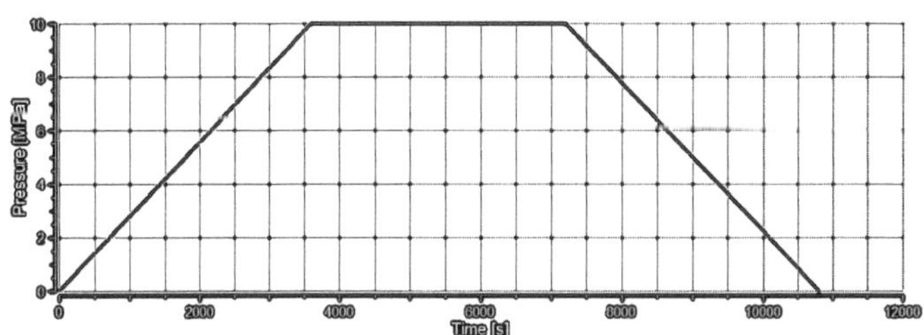

Figure 20 Pressure variation curve for CIC treatment

The results obtained after treatment are shown in the table below:

Table 8 Comparison of the effect of post-processing on stresses and distortions

Final piece	**residual stress(MPa)**	**Distortion (mm)**
Without treatment	1186,85	2,68
With heat treatment (TTH)	289,11	6
With hot isostatic pressing (CIC)	343,02	2,23
With CIC+TTH	89,09	14,93

The heat treatment can be applied before the parts are cut from the SLM manufacturing plate, in order to maintain the geometry during cutting.

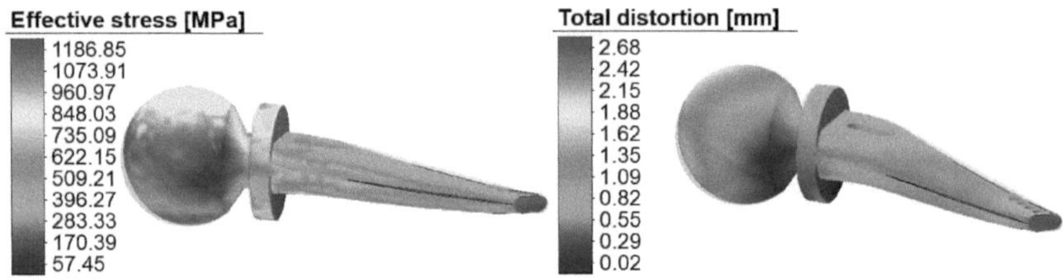

Figure 21 Stress distribution (a) and distortion (b) of a part manufactured without post-processing.

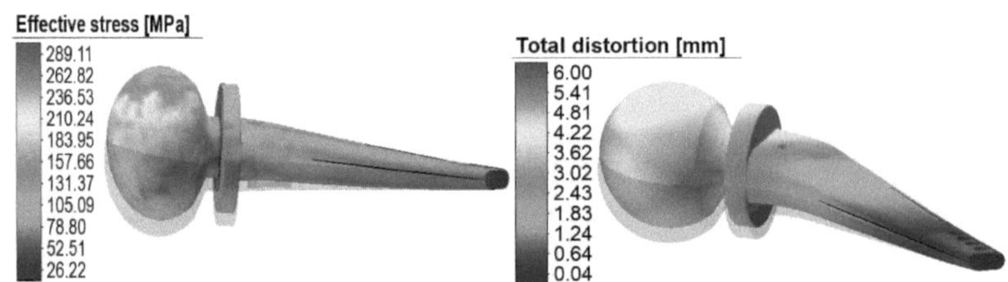

Figure 22 Stress distribution (a) and distortion (b) of an annealed part.

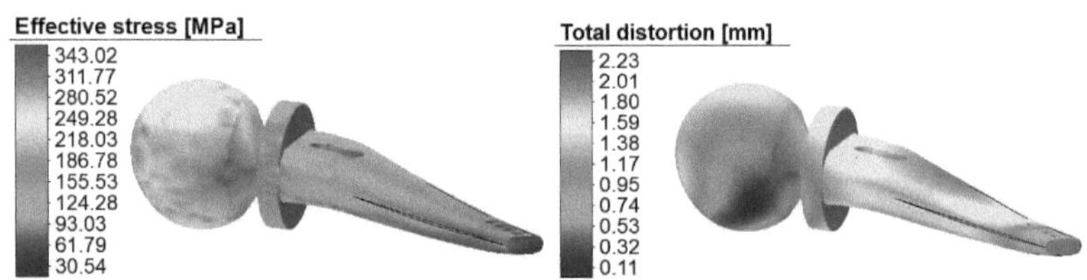

Figure 23 Stress distribution (a) and distortion (b) of a part manufactured with CIC.

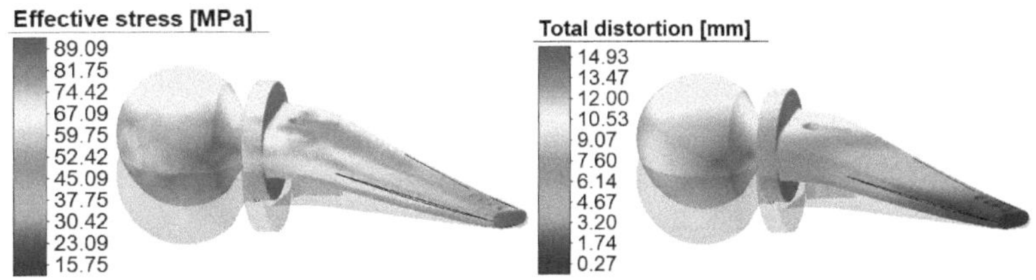

Figure 24 Stress distribution (a) and distortion(b) of a part with CIC + Annealing

Looking at the figures above, we can clearly see the effect of post-processing on the stresses and distortion in the part.

- The maximum stress dropped too much after heat treatment, but distortion increased due to thermal cycling.
- The application of the CIC treatment reduces the values of stresses and distortions, and the values obtained by this treatment are more acceptable.
- A combination of the two treatments reduced the stress values, but the distortion increased remarkably.

4 Conclusion

In this chapter, the results of numerical simulations carried out using Simufact additive software show that the stresses induced on the part at the end of construction are higher at the level of the supports fixed to the plate. These results give us a better understanding of the thermal and mechanical phenomena involved, and their influence on the quality of the parts produced. For example, after the plate has been cut and the supports removed, the maximum stress value equals 1186.85 MPa, concentrating at the head of the rod. However, the deformation increases from one stage to the next until we obtain a final part with a final maximum value equal to 2.68 mm. We will use these values as the basis for the parametric study and optimisation in the next chapter, in order to improve the operating conditions and enhance the quality of the parts produced.

CHAPTER 3: PARAMETRIC OPTIMISATION OF THE SLM PROCESS

CHAPTER 3: PARAMETRIC OPTIMISATION OF THE SLM PROCESS

1 Study of process parameters

Knowing that the LBM process has more than 300 variables that influence the final result of the manufactured parts, our objective is to evolve just a few parameters of the manufacturing machine, in order to find those that have the greatest effect and to find the optimum that minimises the field of stresses and distortions during manufacture.

1.1 Experience plan

We are going to set up an experimental design 2^3 (three factors at two levels). The factors are defined as follows:

- ✓ Powder layer thickness $\mathbf{E_{couche}}$
- ✓ Laser spot power $\mathbf{P_{laser}}$
- ✓ Laser scanning speed $\mathbf{V_{scan}}$

The number of simulations to be carried out is equal to 8 simulations. The possible combinations are shown in the tables below. For each factor we associate the following two levels:

Table 9 Factor table

FACTOR	Low level (-1)	High level (+1)
Layer (µm)	30	60
P_{laser} (W)	100	200
V_{scan} (m/s)	1	2

Simulation number	Ecouche	P_{laser}	VScan
Simulation01	-1	-1	-1
Simulation02	-1	-1	+1
Simulation03	-1	+1	-1
Simulation04	-1	+1	+1
Simulation05	+1	-1	-1
Simulation06	+1	-1	+1
Simulation07	+1	+1	-1
Simulation08	+1	+1	+1

After carrying out the necessary simulations, the results extracted from the Simufact Additive software are shown in the table below.

Table 10 Experience table

Simulation number	Ecouche	P_{laser}	VScan	Residual Max	Max Distortion
Simulation01	30	100	1	1289,75	3.23
Simulation02	30	100	2	1193	2.60
Simulation03	30	200	1	1304,24	6.30
Simulation04	30	200	2	1252,43	5.10
Simulation05	60	100	1	1210,95	3.14
Simulation06	60	100	2	1194,55	2.46
Simulation07	60	200	1	1302,91	6.10
Simulation08	60	200	2	1254,15	4.87

1.2 Interpretation of experimental design results

The next step is to study and determine the most influential factors on the stress and distortion results using static tools such as the response graph and the analysis of variance techniques developed in the Minitab® statistical software.

The pareto diagram shows that laser power and scanning speed have the greatest effect on stresses in relation to layer thickness.

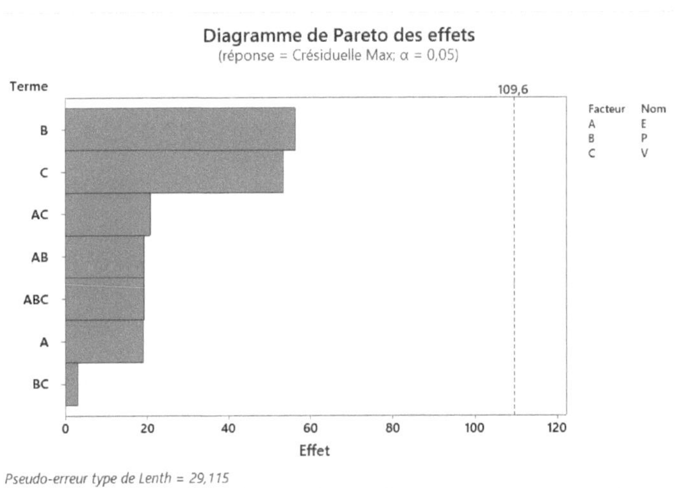

Figure 25Pareto diagram of effects for the constraint

The plots of the main effects shown in Figure 26 show a slight decrease in residual stress caused by the variation in the thickness of the layer from low level to high level, a greater decrease when the velocity is equal to 2 m/s (high level), but an increase in stress values from 100 W to 200 W.

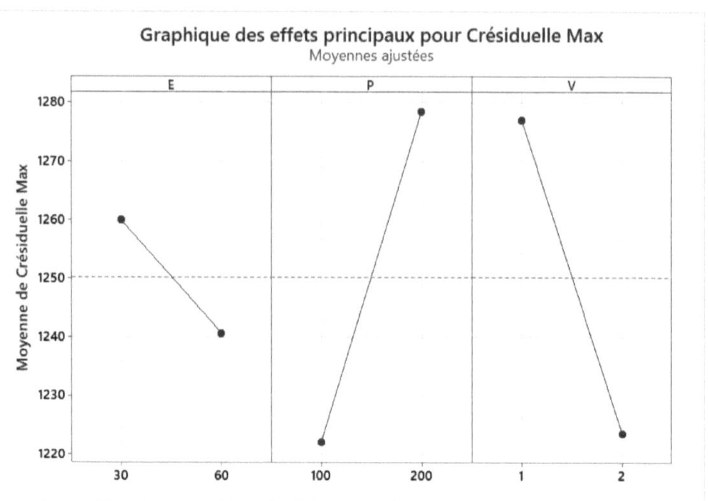

Figure 26 Main effects graph for the constraint

From the interaction diagrams we conclude that there is no interaction between power and speed, and a weak, insignificant interaction between the two pairs Thickness-Speed and Thickness-Power.

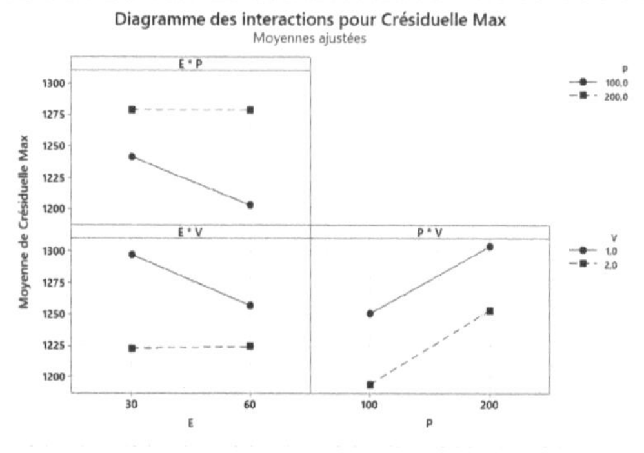

Figure 27Interaction diagram for Max Residual stress

In the contour plot of Stress and V; P, we can see that the stress varies from 1200MPa to 1300 MPa and we can deduce that the minimum residual stress response is obtained for the minimum power (100W) , the maximum layer thickness (60 μm) and the maximum scan speed (2m/s) - See Figures below.

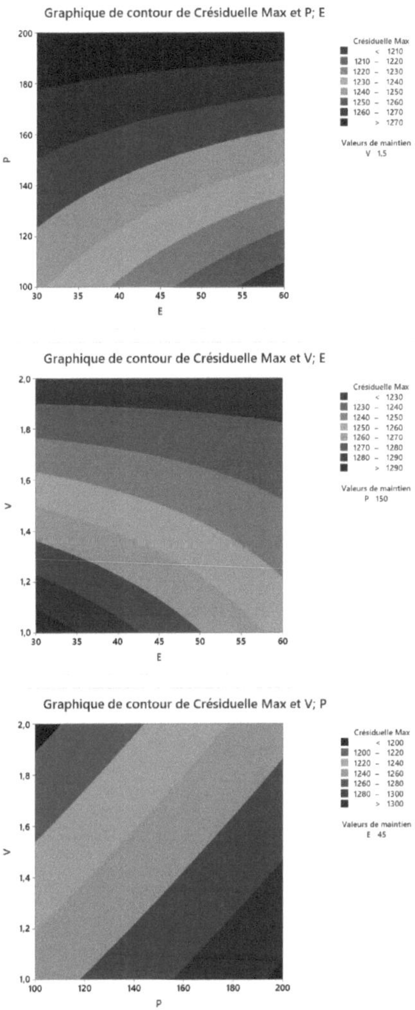

Figure 28: Crésiduelle contour graph

Stress regression equation

Residual Max = 1731 - 10.46 E - 1.852 P - 299.3 V + 0.05159 E*P + 5.255 E*V + 1.222 P*V - 0.02577 E*P*V

The pareto diagram of distortions shows that it depends essentially on the power, which is far greater than the other factors.The scanning speed is also a factor influencing distortions in the room.

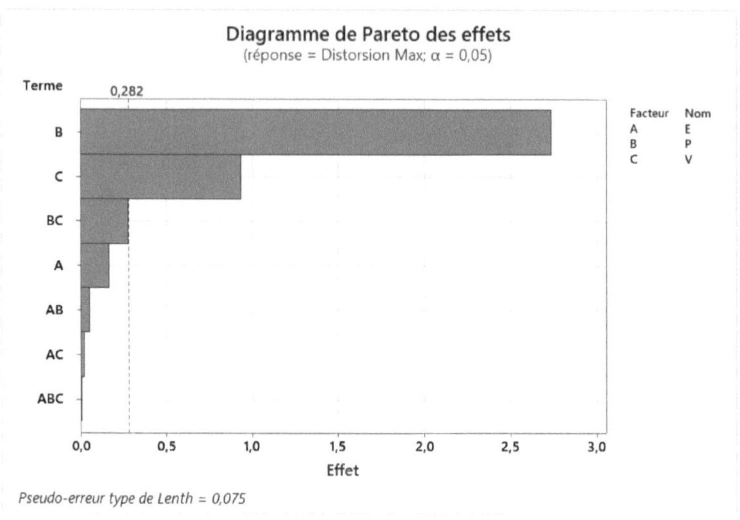

Figure 29: Pareto chart of effects

The main effects plots in Figure 30 show that the variation in layer thickness has almost no effect on the distortion values. These values decrease with decreasing laser spot power. An increase in scanning speed can lead to a decrease in distortion.

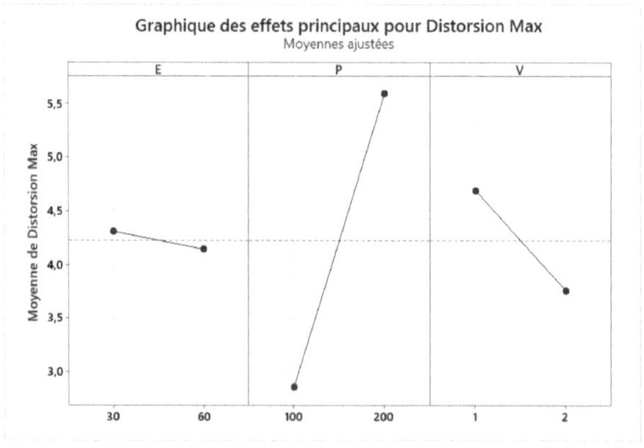

Figure 30: Main effects graph for Distortion Max

The interaction diagrams **in Fig.31** show parallel curves, which means that there is little or no interaction between the different parameters.

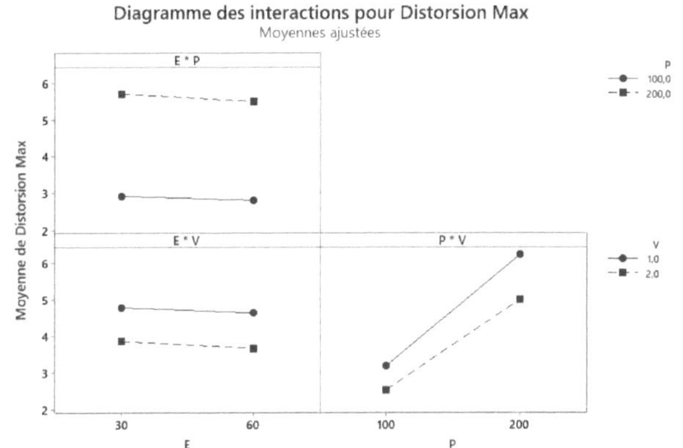

Figure 31: Interaction diagram for Distorsion Max

the contour graph of Max and V distortion; shows a variation in distortion and it can be seen that the lowest values are less than 3 mm obtained for scanning speed values greater than 1.4 m/s and for the minimum laser power (100W).

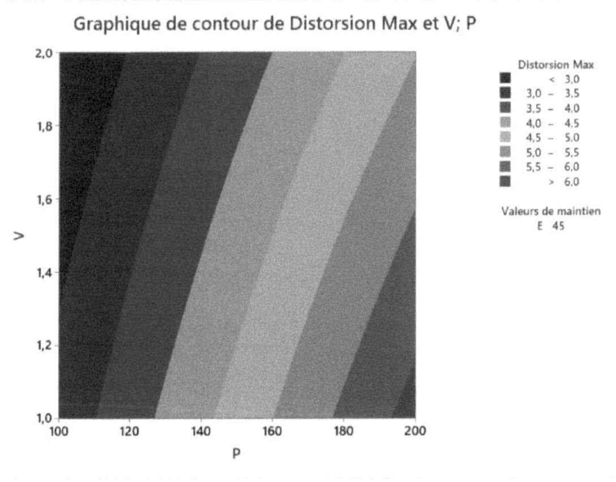

Figure 32: Max Distortion contour graph

Max distortion regression equation in uncoded units

Max Distortion = 0.1300 + 0.003000 E + 0.03770 P + 0.01000 V - 0.000043 E*P - 0.002333 E*V - 0.005900 P*V + 0.000007 E*P*V

2 Results

As mentioned previously, the aim is to define the optimum power/speed/coating thickness based on calculations of the residual stress field and the distortion of the part. To achieve this, the responses were optimised using MiniTab software. Response optimisation allows us to identify the combination of variable parameters that together optimise the response sought.

Table 11: Table of parameters

Response	Objective	Lower	Target	Superior	Weighting	Importance
Max Distortion	Minimum		2,46	6,30	1	1
Residual Max	Minimum		1193,00	1304,24	1	1

Table 12: Variable ranges

Variable	Values
E	30
P	(100; 200)
V	(1; 2)

As shown in **Table 13**, it can be seen that the minimum residual distortion and stress occur for P=100 W, v=2m/s and E= 30 µm, which is therefore the optimum.

Table 13: Solution

Solution	E	P	V	Max. distortion Adjusted value	Residual Max Adjusted value	Composite desirability
1	30	100	2	2,6	1193	0,981602

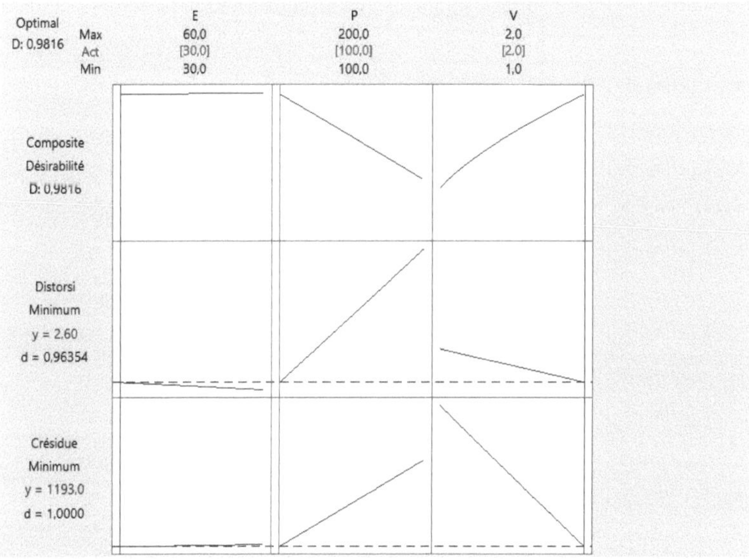

Figure 33: Optimisation diagram

3 Conclusion

The parametric approach makes it possible to identify the optimum configurations that meet performance requirements, while minimising costs and defects. To this end, a numerical experimental design was established. Various influential parameters and their variations were determined. The aim of this experimental design was to analyse the influence of process parameters linked to the machine (layer thickness, laser speed) and linked to the laser (laser power). We confirmed that laser power is the parameter with the greatest influence on stress values, so it needs to be controlled. To create parts with good mechanical properties and good densities, the manufactured part needs to undergo post-treatments that will help to relax stresses and minimise porosity. The effects of these treatments on stresses and distortions have been worked out. Finally, a parametric study was carried out to determine the impact of three manufacturing process parameters on residual stresses and distortions. These parameters are layer thickness, laser power and scanning speed. This study shows that the optimum is achieved for laser power P=100 W, scanning speed v=2m/s and layer thickness E= 30 μm.

CHAPTER 4: TOPOLOGICAL OPTIMISATION OF A KNEE PROSTHESIS FOR ADDITIVE MANUFACTURING

CHAPTER 4: TOPOLOGICAL OPTIMIZATION OF A KNEE PROSTHESIS FOR ADDITIVE MANUFACTURING

1 Topological optimisation for additive manufacturing using SLM

The design of parts developed for additive manufacturing is dealt with in the literature in many different ways. There are examples of design involving topological or parametric optimisation, the emergence of design rules from process characterisation, approaches using specific tools, etc.

Several studies in orthopaedic biomechanics have shown that good prosthesis design through better optimisation of form factors (diameter, thickness, etc.) can improve the lifespan of joint implants, hence the tendency to design prostheses (THRs, TKRs, etc.) with increasingly optimised topology.

Topology optimisation and additive manufacturing are seen as a perfect match, thanks to the ability of FA to produce parts with complex geometries and topologies.

This chapter focuses on the topological optimisation of parts for additive manufacturing. The tools and software used to carry out this type of optimisation will be illustrated through an application on a knee prosthesis using the "ntopology" software. [13]. To facilitate and optimise calculation time, we have assumed that the part is symmetrical.

1.1 Topology-optimised design

In traditional design, an initial digital model of the part (virtual prototype) is tested using various digital simulation calculation tools (mechanical, etc.) to predict its performance. After analysing the calculation results, the geometry of the part is then modified to be digitally re-tested. This process is repeated until a solution is found that meets the requirements of the initial functional specifications.

The efficiency of this design process therefore depends on the designer's ability to come up with appropriate solutions quickly, thereby limiting the number of iterations. This digital design approach often resembles a long trial-and-error process that struggles to converge on the optimum solution. It is therefore becoming necessary to integrate more advanced and easier tools to optimise this design process. Design optimisation or topological optimisation is one of the most widely used numerical methods, especially for additive manufacturing.

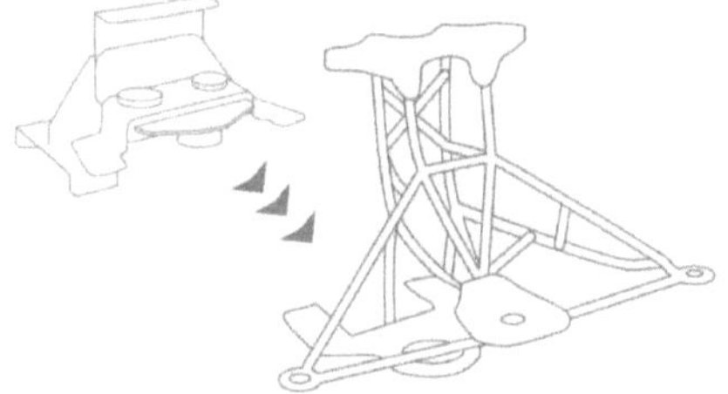

Figure 34: Mounting bracket optimisation example

1.2 Lattice structures

Lattice structures are bio-inspired configurations based on repetitive unit cells composed of strips or lattices with octagonal, honeycomb or random patterns. Engineers should consider using these topologically-optimised structures in additive manufacturing for the following reasons:

These are complex structures that cannot be produced using conventional manufacturing technologies. These structures enable us to save on materials by reducing weight. The property of low mass at high strength is a key advantage of these structures. [14] This type of structures are similar to bone structures fig.35 which allows good bone anchorage (osseointegration) with implantable medical devices. [15] [16]

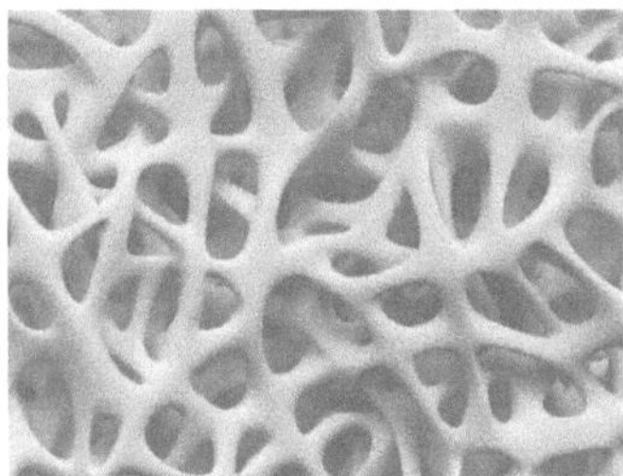

Figure 35: Close-up of the cancellous bone structure

Figure 36: Different types of lattice structures produced by nTopology software

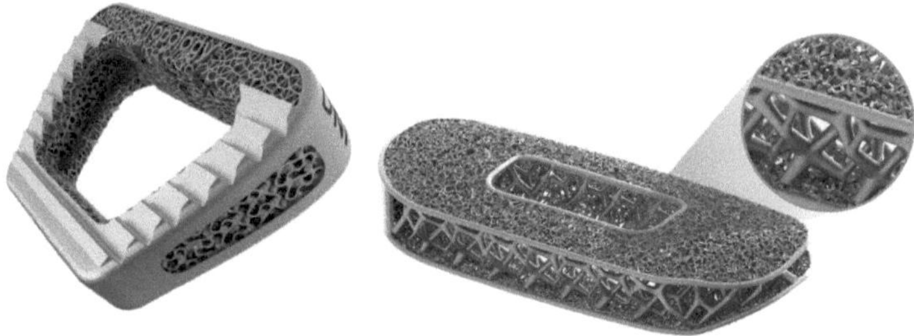

Figure 37: Integration of lattice structures on ALIF implants using nTopology

1.3 Steps in topological optimisation

In order to optimise the topology, the designer must go through a number of important steps after running the simulation and calculation. Most of them follow a general workflow:

a) The first step is to define the design space, also known as the "authorised volume", which may be larger than the part to be redesigned.
b) The second stage is material identification: it is essential to make the necessary selection of materials for each part. Topological optimisation depends on the mechanical properties of the materials, which naturally affect the final optimum geometry.
c) The next step is to apply geometric restrictions: this can be interpreted as setting the areas that are not allowed to change during the optimisation process. In technical terms,

these regions are called frozen regions. The solver will leave these regions as they are while removing the surrounding material to find the optimal topology.

d) The next step is to identify the mechanical stresses applied: all the mechanical loads perceived by the part must be taken into account to obtain the best possible result, as well as identifying any accidental stresses (e.g. shocks).

e) Then apply the constraints and the objective function defined by the designer to offer several part design options. In this way, the solver can continue to remove unnecessary elements until it finds the optimum topology.

f) The final stage, carried out by the designer, consists of reconstructing the part based on the 'ideal' geometry - functionally optimal but often geometrically very complex (numerous nested hollow areas) - and transforming it into a 'manufacturable form' adapted to the various constraints of the transformation process in particular.

Figure 38: Principle of topological optimisation

The result of topological optimisation obtained by digital software can sometimes be triangular and of complex shape and geometry that conventional machining processes are unable to achieve. This makes it easy to manufacture these parts using additive manufacturing processes.

This innovative tool saves material and manufacturing time, and improves and accelerates the design of very high-performance products. It also makes it possible to minimise compliance and mass. [17]. Topological optimisation can also be put to good use as part of a "classic" redesign, resulting in new, improved geometries that replace the initial design.

Indeed, the use of topological optimisation tools can suggest subversive shapes that the designer would not naturally have thought of.

2 Case study: Topological optimisation of knee prosthesis

2.1 General working methodology

This part is divided into three stages:

- The first step is to apply the static analysis to the implant before topological optimisation.
- We then apply topological optimisation using lattice structures in a second step.
- The third stage involves validation using static analysis.

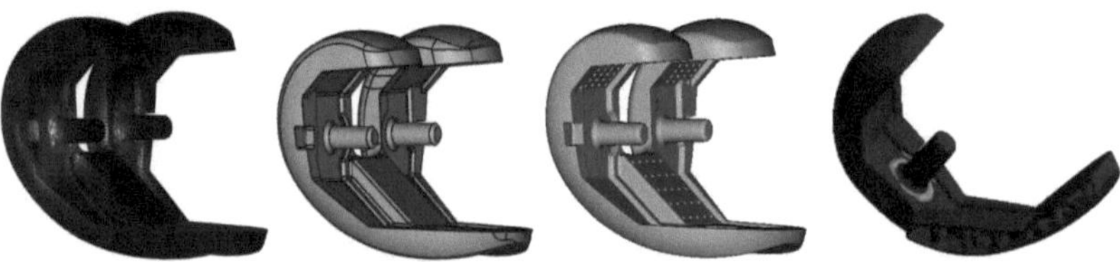

Figure 39: General workflow

2.2 Choice of software

CAD software does not offer this topological optimisation function, so some publishers have developed dedicated software. One of the pioneers is undoubtedly Altair with its very first OptiStruct solution, later joined by another solution, Altair Inspire. Others include Ansys, Dassault Systèmes, Autodesk and nTopology.

For topological optimisation we chose nTopology software, which is an advanced, fast-growing, revolutionary engineering software, its aim being to create a better system for product development based on data and physics, trying to create the next generation of engineering design tools for advanced manufacturing, including additive manufacturing. The most important aim of this choice of software is that nTopology is essentially designed, as its name suggests, for topological optimisation using different part lightening methods (lattice structures, perforation etc.) plus it integrates design, simulation and manufacturing tools in a single environment, capturing engineering knowledge and enabling automation along the way.

[APPENDIX 2]

Static analyses are carried out using ANSYS software.

2.3 Geometric modelling

A knee prosthesis consists mainly of 3 components: the tibia prosthesis, the femur prosthesis and an insert.

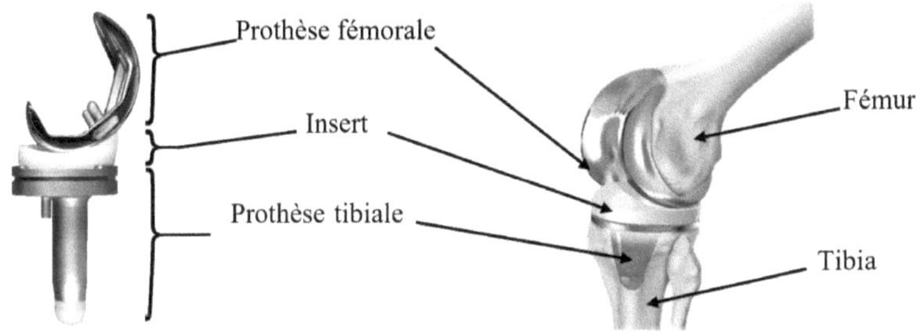

Figure 40The components of a total knee prosthesis

The geometric CAD model designed by CATIA V5 was imported from GRABCAD. As the part is symmetrical, we decided to apply the static analysis before and after optimisation to only half of it.

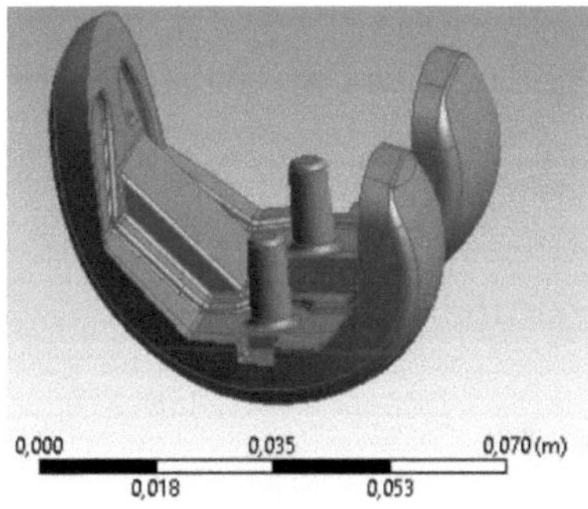

Figure 41Generation of the geometric model of a femoral component of a knee prosthesis

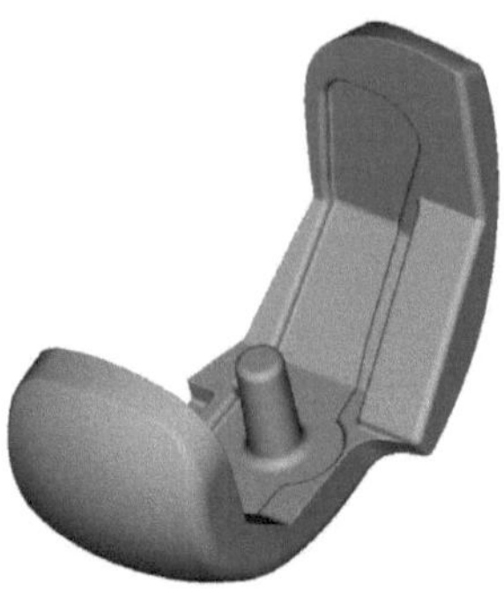

Figure 42: Half of the part

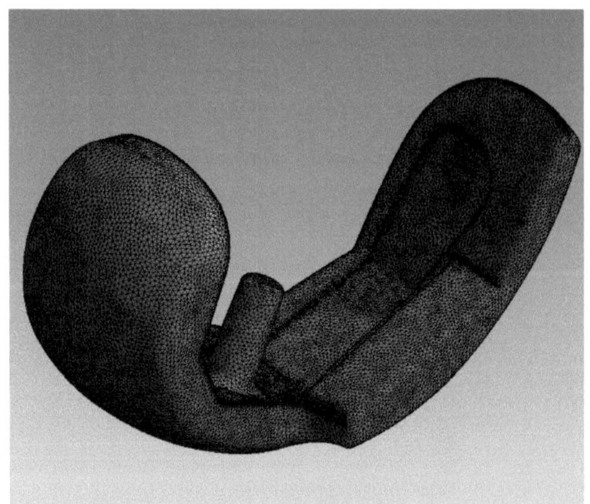

Figure 43CAD model mesh of the femoral component

Table 14 Mesh results

Type	Triangular
Number of nodes	208699
Number of elements	1185375

2.4 Loading and limit conditions

The knee joint is a complex structure in the human body which is subjected to a critical load simultaneously when performing different physical activities such as walking, running, rotating, sitting, static positions, etc...

Studies have consistently estimated the maximum compressive forces on the knee joint to be 4 to 4.5 times body weight during daily activities [18].

The cylindrical forms shown in the figure with blue colour are fixed by adhesion to the upper part of the knee, and that the force applied on both surfaces (red colour) is equal to 2200 along the z axis, (weight = 50kg).

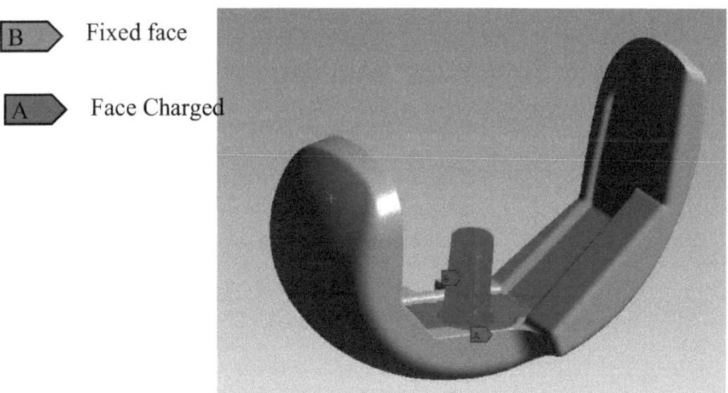

Figure 44Forces and boundary conditions applied to one half of the femoral implant (ANSYS)

2.5 nTopology workflow

Topological optimisation using lattice structures in the nTopology software is carried out as follows:

- **Convert the imported part into an implicit body**

After importing the CAD part, we convert it into the implicit format, which is a uniquely powerful and lightweight way of defining, representing and changing three-dimensional objects using a simple mathematical function to represent the solid.

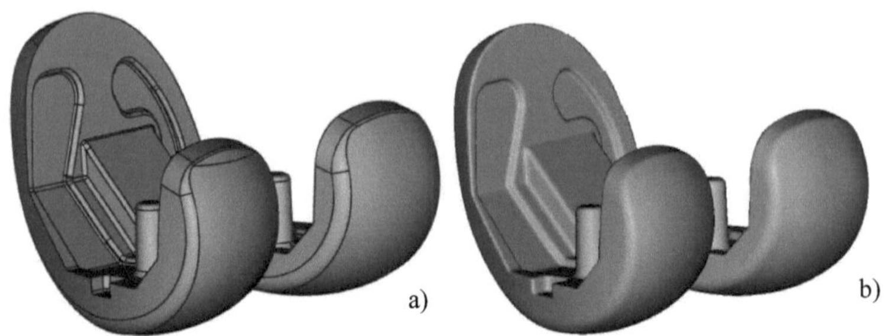

Figure 45 Geometric model of the implant: a) Format explicit,b) Format Implicit

- **Internal volume reduction using lattice structures**

In nTopology, the application of complex structures in general is fairly easy and practical. In our case, we worked with a block called "Shell and Volume lattice", which consists of deconstructing an implicit body and filling it with volume lattices.

Table 15: Parameter for implant lightening

Body	The Implicit Body		
Type of lath	cubic centred on the body		
Cell size (mm)	10	10	10
Rotation	0	0	0
Lath thickness (mm)	1		
Shell thickness (mm)	1		
Mixing radius (mm)	0.5		

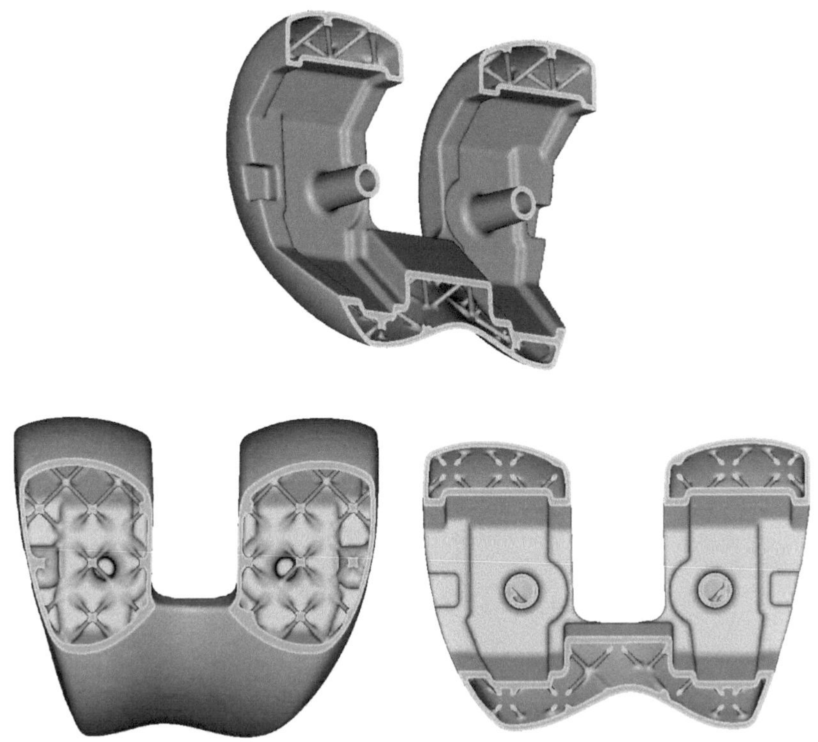

Figure 46: Cross-section after internal lightening

After this operation, the new weight of the part will be 61.7971. These lattices have enabled us to lighten it by more than half.

- **Creation of powder discharge events**

In order to really save weight, we need to be able to remove the powder from the material remaining inside the part after it has been 3D printed by laser fusion on a powder bed, otherwise it will be filled with unfused powder. To achieve this, we chose to make perforations on the faces of the part to guide the remaining material. Fig.59

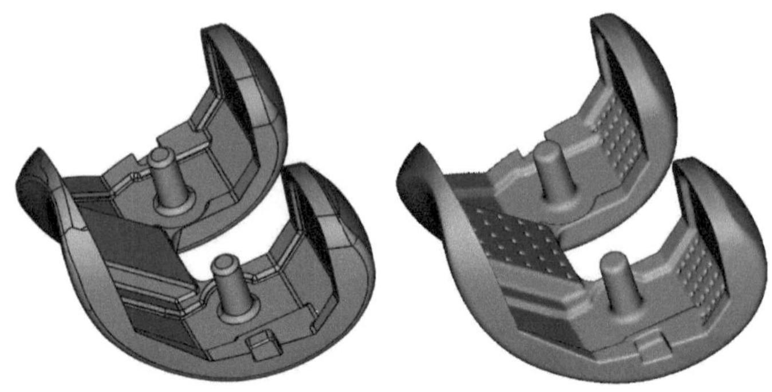

Figure 47: Part after perforation

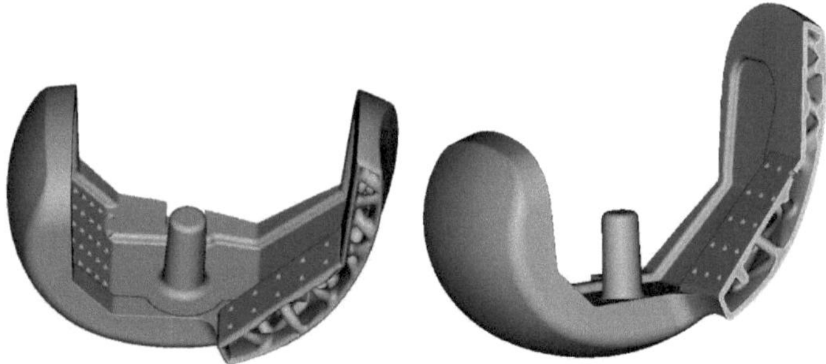

Figure 48: Half of the part to be exported for analysis

- **Part export: EF mesh export with sets**

Exporting a part with a complex geometry and a lot of detail in CAD format (STEP or Parasolid) is one of the most important steps. The steps followed are :

1. Choose the faces on which the force and boundary condition are applied
2. Generate an FE(finite element) mesh for the model.
3. Use the **Export EF mesh** block.
4. Enter the mesh and add the limits in the **sets** field.
5. Update the mesh export path.
6. Select the file type as **Ansys Mechanical Input (.cdb)**

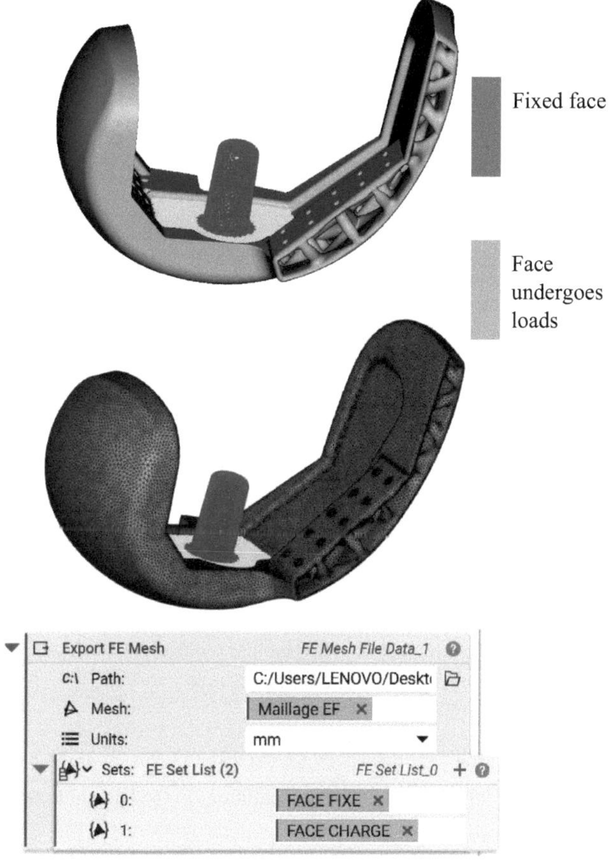

Figure 49: Export steps for the optimised part

- **Validation of the part after topological optimisation**

After exporting the part optimised by nTopology in Ansys Mechanical Input format (.cdb), it must now be imported into ANSYS WORKBENCH using the "external model" tool.

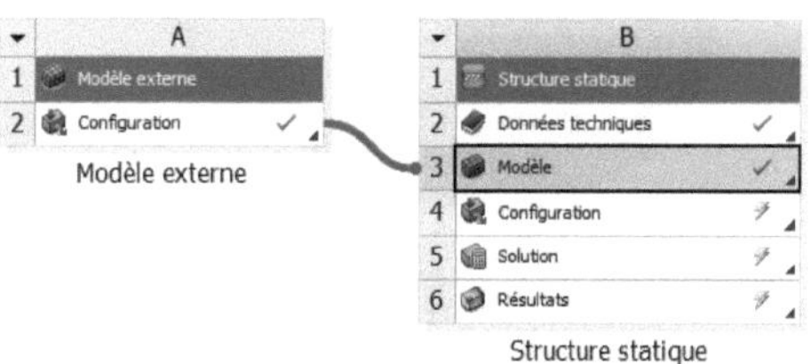

Then click on **configuration** to modify the properties in the external model schema (Uncheck the validity check of the locked CDB file).**Fig.63**

Propriétés de Fichier - C:\Users\LENOVO\Desktop\final\nTop\fe mesh with sets FORCE BC.cdb

	A	B	C
1	Propriété	Valeur	Unité
2	⊟ Description		
3	Source d'application	MAPDL	
4	⊟ Définition		
5	Système d'unité	Metric (kg,m,s,°C,A,N,V)	
6	Traitement des composants nodaux	☑	
7	Clé de composant nodal		
8	Traitement des composants d'éléments	☑	
9	Clé de composant d'élément		
10	Traiter les composants de faces	☑	
11	Clé de composant de face		
12	Traitement des données du modèle	☑	
13	Traitement des éléments Mesh200	☑	
14	Vérifier la validité du fichier CDB bloqué.	☐	
15	Méthode de renumérotation des nœuds et des éléments	Automatic	
16	Type de transformation	Rotation et translation	

Figure 50: External model configuration

Open the model in static structure and update it to see that the mesh and sets are imported as named selections.

Apply boundary conditions to selections named using Direct FE in the Environment ribbon. Note that node sets only support boundary conditions in the Direct FE drop-down list.

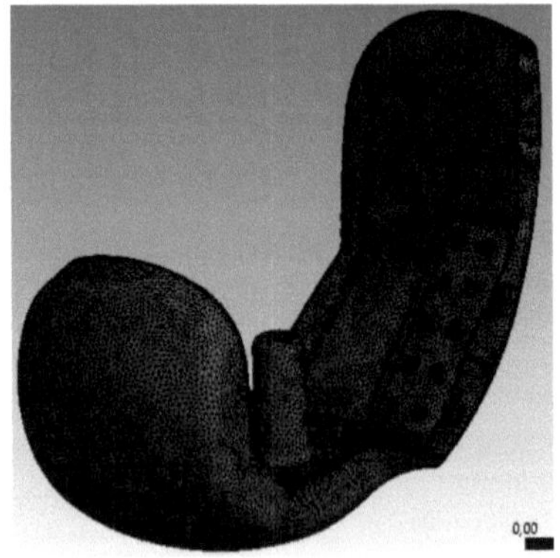

Figure 51: Part mesh

Table 16: Mesh results in Ansys

Number of nodes	285864
Number of elements	1282327

B Fixed face

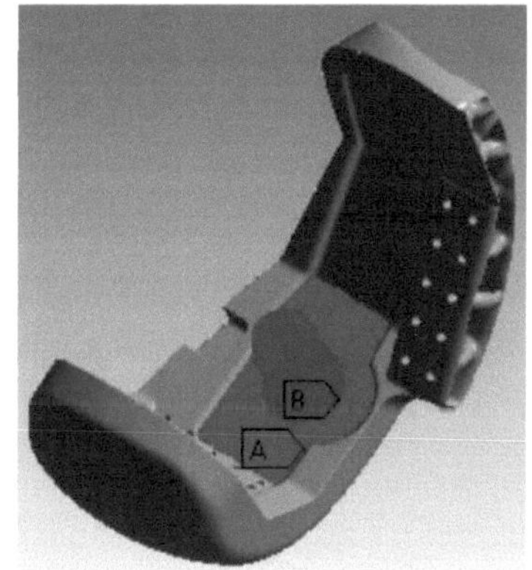

Figure 52: Application of forces and CL

We can now resolve the EF configuration and add the plots we want to view.

3 Results

3.1 Before topological optimisation

After applying the previous steps, the results obtained by ANSYS are shown below.

- **<u>Displacement in mm :</u>**

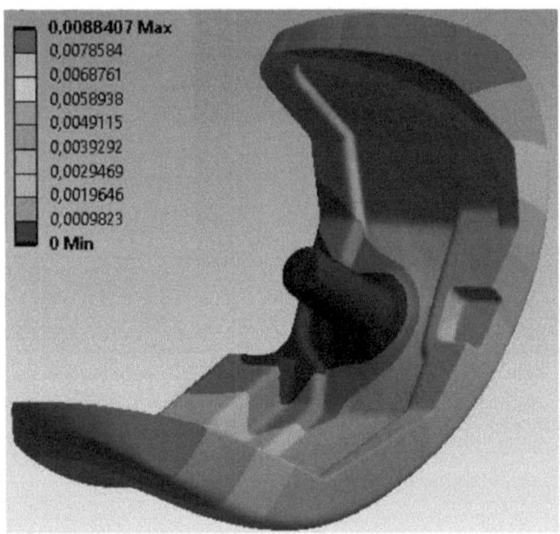

- **<u>Equivalent stress (Von Mises) in MPa:</u>**

➢ **Deformation :**

3.2 After optimisation

➢ **Displacement in mm :**

➢ **Equivalent stresses (Von Mises) in MPa :**

➢ **Deformation :**

The results of the static analysis can be summarised in the table below

Table 17: Comparison of results before and after optimisation

		Part before optimisation	Part after optimisation
Von Mises(MPA)	Max	74,572	448,29
	Min	$6{,}7076e^{-5}$	$5{,}1048e^{-5}$
Deformation	Max	$9{,}2409.\,e^{-4}$	$4{,}3619.\,e^{-3}$
	Min	$6\,{,}3037.\,e^{-10}$	$8{,}8305.\,e^{-10}$
Total displacement Max(mm)		$8{,}8407.\,e^{-3}$	$5{,}1446.\,e^{-2}$
Total weight(g)		154,3271	77,95

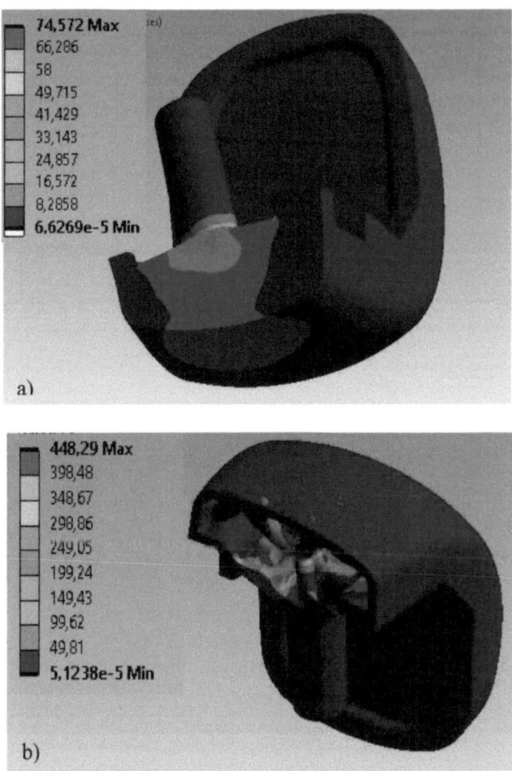

Figure 53: Stress distribution inside the part a) initial, b) optimised

Initially, the same loads and boundary conditions are applied to one half of the femoral component of the knee prosthesis, as represented in Figure (63-71) by the colours red and blue respectively.

By observing the stress distribution, we can say that the areas affected in the part with the lattice structures have decreased compared with those before optimisation, especially on the underside of the part. However, the maximum stress values have increased significantly, as shown in Table 19.

4 Conclusion

The main aim of optimising this femoral component is to save material while maintaining Von Mises stress values below the yield strength of Ti6Al4V titanium, which is equal to

863MPa [19] so that the part remains within the safe elastic zone. For our femoral part, we have gained almost half its initial weight with a maximum value of Von Mises below the material's elastic limit, which means that the part is in the safe elastic zone. As a result, the new optimised part is well made.

CONCLUSION AND OUTLOOK

Finally, the research presented in this dissertation has enabled us to discover the subject of metal additive manufacturing, in particular the process of laser fusion on a powder bed known as LBM / SLM. This process has a number of advantages, such as geometric flexibility, but also some disadvantages, such as the presence of residual stresses, the effects of which must be reduced by optimising the 3D printing parameters or by using post-treatment processes.

Good control of the process is important in order to increase the quality of the parts manufactured, which is why we carried out a digital simulation of the metal additive manufacturing process using SIMUFACT ADDITIVE software. This simulation was carried out on a medical implant, a hip prosthesis, manufactured from TI-6Al-4V titanium superalloy. The simulation results enabled us to quantify the levels of residual stress and strain at the end of the laser-melting phase of the metal particles.

The influence of process parameters on the final state of the printed part was investigated using a study based on the design of experiments method. This optimisation provides information that can be used not only for a better selection of manufacturing parameters, but also for the elimination of costly post-processing problems and challenges. Certainly, the results of this research will help users to improve their experience of the LBM/SLM process, or similar procedures, through practical applications.

The application of the metal additive manufacturing process in the printing of implantable medical devices offers an advantage in the manufacture of customised implants for patients with lightened component parts following structural or topological optimisation. This structural optimisation using lattice structures was applied to a femoral component of a knee prosthesis.

With the aim of improving performance and reducing distortions in the SLM process, there are several areas of research and optimisation worth exploring.

For example, simulation of the metal additive manufacturing process includes a fair number of thermomechanical phenomena and metallurgical transformations. A study of these phenomena is a very interesting way of understanding the process.

From a material point of view, the modelling we have created does not take into account the enthalpy of fusion during the solid/liquid transition of the powder at the time of laser fusion (software at the microscopic scale), which allows us to take into account a wide range of scales necessary for the realism of the modelling.

As another perspective, the topological optimisation of medical implants must be further explored by using surface lattice structures similar to the internal tissues of bone in order to approximate the Young's modulus of the latter. This optimisation may have consequences for the morphology of surfaces in direct contact with the bone in order to achieve successful osseointegration.

From a numerical point of view, it would be very interesting to carry out an analysis of the polycyclic fatigue behaviour of the optimised parts using the Ncode DesignLife software in order to predict the lifespan of these components and the level of damage during the application of these medical implants.

REFERENCES

[1] Horvath J., "A Brief History of 3D Printing. Mastering 3D Printing," 2014.

[2] W. H. charles, "Apparatus for Production of Three-Dimensional Objects by Stereolithography". U.S. Patent 4,575,330, 1986.

[3] MR B. ,. E. F. ,. S. G. Ali Foroozmehr, "Finite Element Simulation of Selective Laser Melting process considering Optical Penetration Depth of laser in powder bed," *SCIENCES DIRECT,* 2015.

[4] "3Dnatives,, [Online]. Available: https://www.3dnatives.com/titane-impression-3d-ti6al4v-09022021/#.

[5] R. P. v. Julie, "Determination of the thermal properties of Ti-6Al-4V alloy for the study of adiabatic shear," in *Congrés Français de Thermique*, 2016.

[6] "MatWeb Titanium alloys," [Online]. Available: http://www.matweb.com/search/datasheet.aspx?bassnum=MT0001&ckck=1.

[7] L. L.Van Belle, "Analysis, modelling and simulation of stress onset in metallic laser melting," *Hal,* 2013.

[8] A. AGOUZOUL, "Nouvelles méthodes numériques pour la simulation de l'impression 3D métallique," 2020.

[9] CETIM, Metal additive manufacturing - The fundamentals, CENTRE TECHNIQUE DES INDUSTRIES MÉCANIQUES (Cetim), 2019

[10] G. Lefevre, "Simufact Additive simulation software," A3DM Magazine, [Online]. Available: https://www.a3dm-magazine.fr/news/logiciels/logiciel-simulation-simufact-additive.

[11] I. A. Roberts, INVESTIGATION OF RESIDUAL STRESSES IN THE LASER MELTING OF METAL POWDERS IN ADDITIVE LAYER MANUFACTURING, 2012.

[12] V. CHASTAND, "Etude du comportement mécanique et des mécanismes d'endommagement de pièces métalliques réalisées par fabrication additive," HAL , 2016.

[13] deloitte, "challenges of additive manufacturing," 2019.

[14] a. h. azman, Method for integration of lattice structures in design for additive manufacturing, HAL , 2017 .

[15] P. RENARD, "Additive manufacturing: towards more efficient titanium-ceramic implants?" 2018. [Online]. Available: https://www.devicemed.fr/dossiers/materiaux/metaux/fabrication-additive-vers-des-implants-plus-efficaces-en-titane-ceramique/15075.

[16] A. C. ,. D. G. ,. H. j. ,. C. S. C. Emanuel Sachs, "Surface Texture by 3D Printing".

[17] Y. Louvigny, J. Nzisabira, B. Meunier and P. Duysinx, "Topological optimisation and additive manufacturing,improving the design chain," 2015.

[18] I. M. VENKATA PAVAN KUMAR, "DESIGN AND STRUCTURAL ANALYSIS OF KNEE IMPLANTS USING DIFFERENT MATERIALS," *INTERNATIONAL JOURNAL OF ADVANCE SCIENTIFIC RESEARCH AND ENGINEERING TRENDS,* 2020.

[19] R. Sepe, S. Franchitti, R. Borrelli, F. Di Caprio, E. Armentani and F. Caputo, "Correlation between real geometry and tensile mechanical behaviour for Ti6Al4V electron beam melted thin specimens," 2020.

[20] R. Ponche, "Design methodology for additive manufacturing, application to powder projection," 2013.

[21] "Principle of additive manufacturing by laser fusion," [Online]. Available: https://www.a3dm-magazine.fr/.

[22] G. Lefevre, "The fundamentals of metal additive manufacturing," 2021. [Online]. Available: https://www.a3dm-magazine.fr/.

[23] "what is additive manufacturing fa," [Online]. Available: https://www.farinia.com/fr/blog/quest-ce-que-la-fabrication-additive-fa.

[24] F. S. D. by Floriane LAVERNE, "Additive manufacturing -general principles," *Techniques de l'Ingénieur ,* 2016.

[25] Aubert&Duval, "TA6V," [Online]. Available: https://www.aubertduval.com/results/.

[26] "Heat treatment of titanium and titanium alloys," [Online]. Available: https://www.bodycote.com/.

[27] S. MISCHLER, "Tribologie des prothèses de hanche," *Techniques de l'ingénieur ,* 2013 .

[28] T. D. Health, "hip prosthesis," 2021. [Online]. Available: https://www.tunisiedestinationsante.com/prothese-de-la-hanche/.

[29] "Lambert clinic hip anatomy," 2016. [Online]. Available: http://cliniquelambertphysio.com/conseils/hanche.

[30] "orthosud anatomy of the hip," 2016. [Online]. Available: http://www.orthosudmontpellier.com/les-pathologies/hanche/anatomie-de-la-hanche.html.

[31] P. B. J. Castaing, Anatomie fonctionnelle de l'appareil locomoteur, Vigot, 1996.

[32] D. Kassab, "The Hip," 2021. [Online]. Available: http://www.orthokassab.com/chirurgie-prothese-totale-de-hanche-paris-traitement-arthrose-hanche-tunisie/.

[33] B. Vayre, "Design for additive manufacturing, application to EBM technology," *HAL* , 2016 .

[34] A. B. ,. H. B. D. R. Ruyssen et al, "Contribution to the modelling of the Additive Manufacturing process: laser melting on powder bed," *HAL,* 2017.

[35] G. B. e. a. Benedek, "Physics with illustrative examples from medicine and biology," 2000.

[36] A. B. B. S. Gaşmi B. et al, "TRIDIMENSIONAL FINITE ELEMENT ANALYSIS OF STRESSES IN TOTAL HIP PROSTHESES," *ResearchGate,* 2015.

[37] D. Cattan, "Arthroscopic techniques. Prostheses," 2008. [Online]. Available: http://www.arthroscopie.fr/hanche/procedure-chirurgicale. html.

[38] A. Hichem, "Bio functionality of a pair of HDPE/ceramic biomaterials " numerical application to the head, cup/cotyl association "," 2019.

[39] K. M'hamed, "Simulation Numérique par la Méthode des Eléments Finis du Comportement Mécanique d'une Prothèse Totale de Hanche Fabriquée dans un Alliage à base de (Cr-Ni-Mo)," 2005.

[40] L. W. ,. B. H. H. e. B. H. M. HAMZA Samir et al, "DIGITAL SIMULATION OF JOINT PROSTHESES (HIP, ANKLE AND SHOULDER)," in *Application médicales de l'informatique nouvelle approche* , 2010.

[41] "Laser metal Fusion," 2021. [Online]. Available: https://www.trumpf.com/en_CA/solutions/applications/additive-manufacturing/laser-metal-fusion/.

[42] "Focus on laser metal powder fusion and its applications," [Online]. Available: http://traitementsetmateriaux.fr/Archives-article/Fiche/1639/.

[43] D. A. M. Dr Thierry Rouach, "Update on the concept of osseointegration," MARS 2010. [Online]. Available: https://www.lefildentaire.com/images/stories/articles2/clinic-format.continue-actualisation-concept-osteointegration/clinic-format.continue-actualisation-concept-osteointegration.pdf.

[44] Y. PAULIAT, "L. A. F. Additive, "Ensil mix."," 2013-2014. [Online].

[45] "metal blog," [Online]. Available: https://metalblog.ctif.com/2018/03/05/les-alliages-de-titane-pour-le-medical/.

[46] "3dnatives," [Online]. Available: https://www.google.com/imgres?imgurl=https%3A%2F%2Fwww.3dnatives.com%2Fwp-content%2Fuploads%2Fdentaire.jpg&imgrefurl=https%3A%2F%2Fwww.3dnatives.com%2Ffabrication-additive-dentaire-210220193%2F&tbnid=SMyQvrTXTFAh9M&vet=10CAMQxiAoAGoXChMImMTQlNSM9QIVAAAAAB.

[47] "roseler Blog," [Online]. Available: https://roslerblog.com/2019/10/29/joint-reconstruction-part-3-surface-finishing-standards/.

[48] O. ANSART, "Université du Maine," [Online]. Available: http://perso.univ-lemans.fr/~fcalvay/projetsmnrv/model_crash_abaqus.htm.

[49] A. W. G. a. H. G. Lemu, "A case study on topology optimized design for additive manufacturing," 2017.

[50] H. M. Kamel, "MODELING AND SIMULATION OF A HIP PROSTHESIS IMPLANTATION," 2019.

[51] D. J. Peter St. Clair, "DENTAL MATERIALS 101," JUNE 2020. [Online]. Available: https://www.jpeterstclairdentistry.com/blog/dental-materials-101-2/.

[52] P. RENARD, "Metal additive manufacturing for resolutions below 15 μm," DeviceMed , SEPTEMBER 2020. [Online]. Available: https://www.devicemed.fr/dossiers/sous-traitance-et-services/impression-3d-sous-traitance-et-services/fabrication-additive-metal-pour-des-resolutions-inferieures-a-15-%C2%B5m/24517.

[53] "Industrial applications of Renishaw metal additive manufacturing technology," Renishaw , [Online]. Available: https://www.renishaw.fr/fr/applications-industrielles-de-la-technologie-de-fabrication-additive-metallique-renishaw--15256#:~:text=La%20fabrication%20additive%20m%C3%A9tallique%20est,carburant%20et%20des%20aubes%20directrices..

APPENDICE

APPENDIX 01: LARSON HIP PROSTHESIS

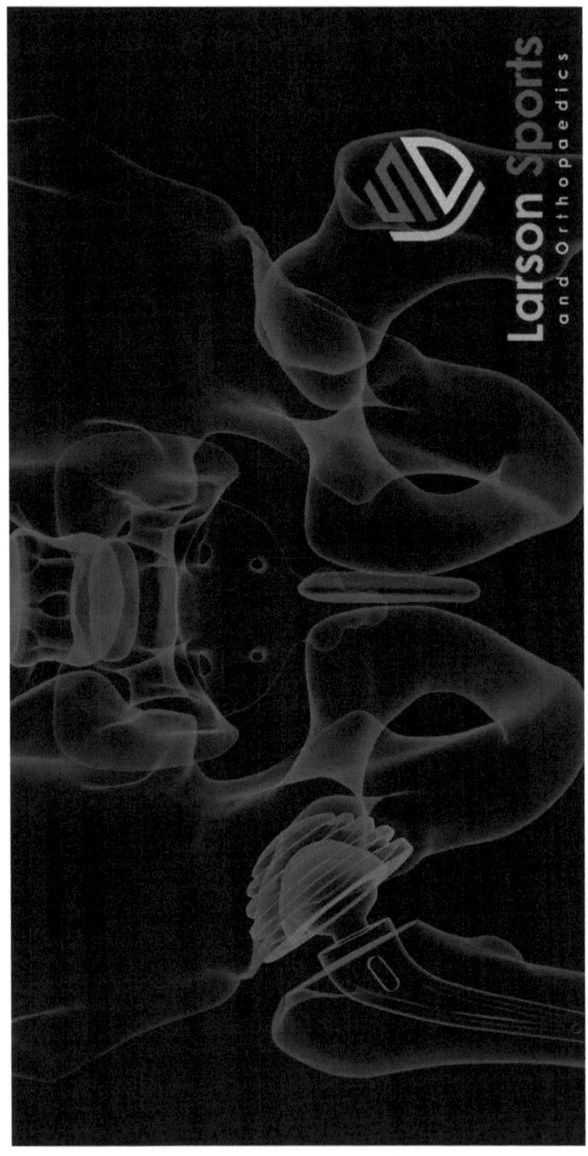

https://larsonsportsortho.com/

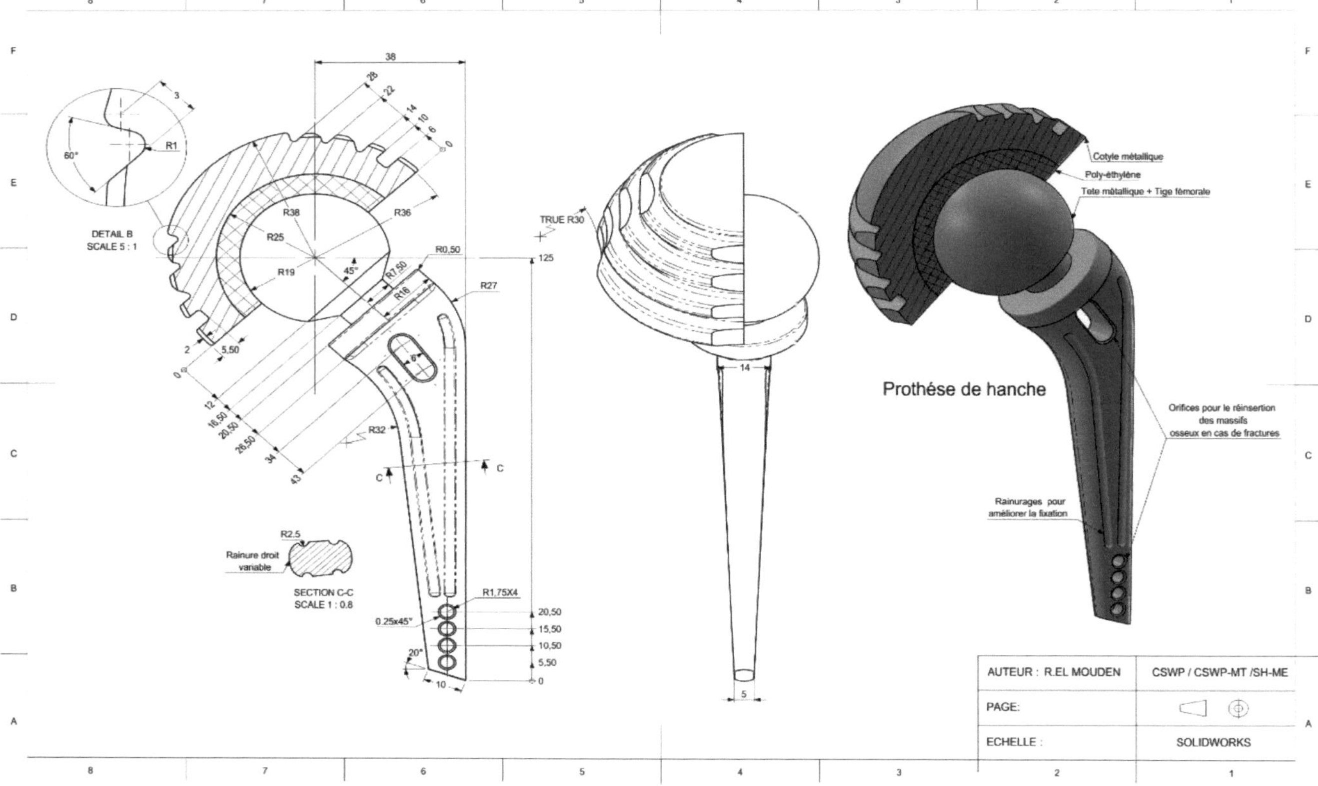
Prothése de hanche
Cotyle métallique
Poly-éthylène
Tete métallique + Tige fémorale
Orifices pour le réinsertion
des massifs
osseux en cas de fractures
Rainurages pour
améliorer la fixation
DETAIL B
SCALE 5 : 1
SECTION C-C
SCALE 1 : 0.8
Rainure droit
variable
TRUE R30
AUTEUR : R.EL MOUDEN
CSWP / CSWP-MT /SH-ME
PAGE:
ECHELLE :
SOLIDWORKS

Appendix 02: KNEE PROTESE

Total knee replacement (TKR) involves replacing damaged parts of the joint with artificial components. The components are made from materials that are particularly resistant to mechanical and abrasive stress. Osteoarthritis is the most common cause of TKR surgery. This pathology wears away or completely destroys the articular cartilage. The result is pain when walking or even at rest, which reduces quality of life. The aim of this operation is to reduce this pain and improve the mobility and stability of the knee.

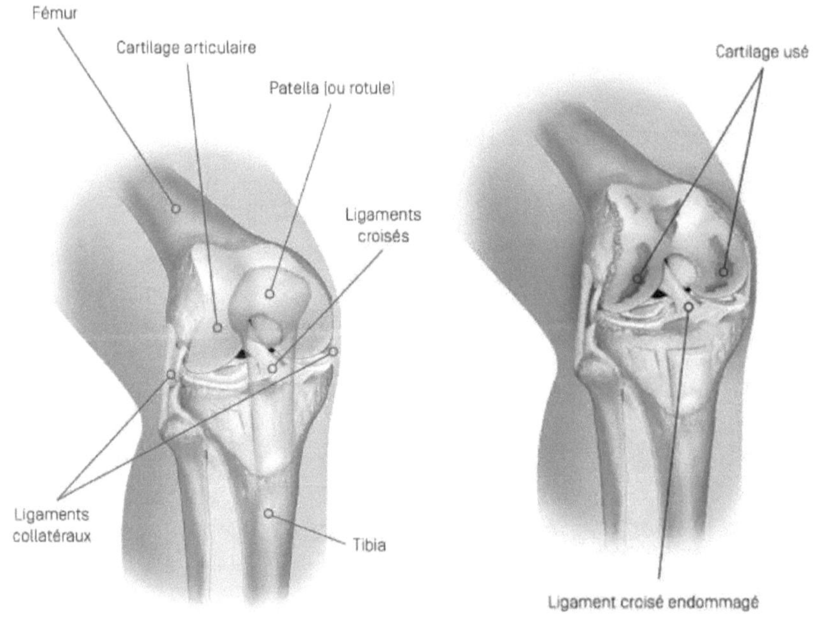

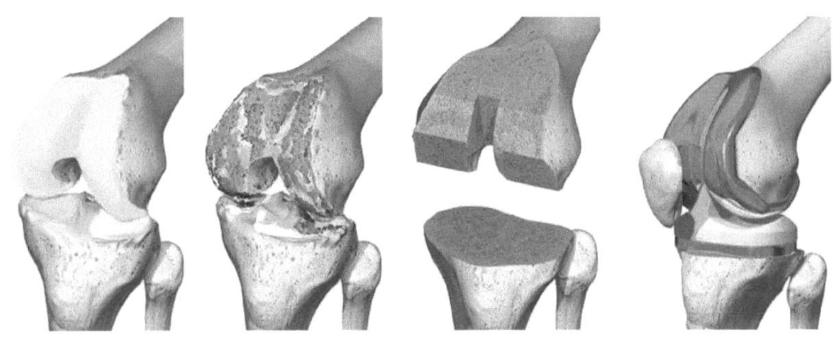

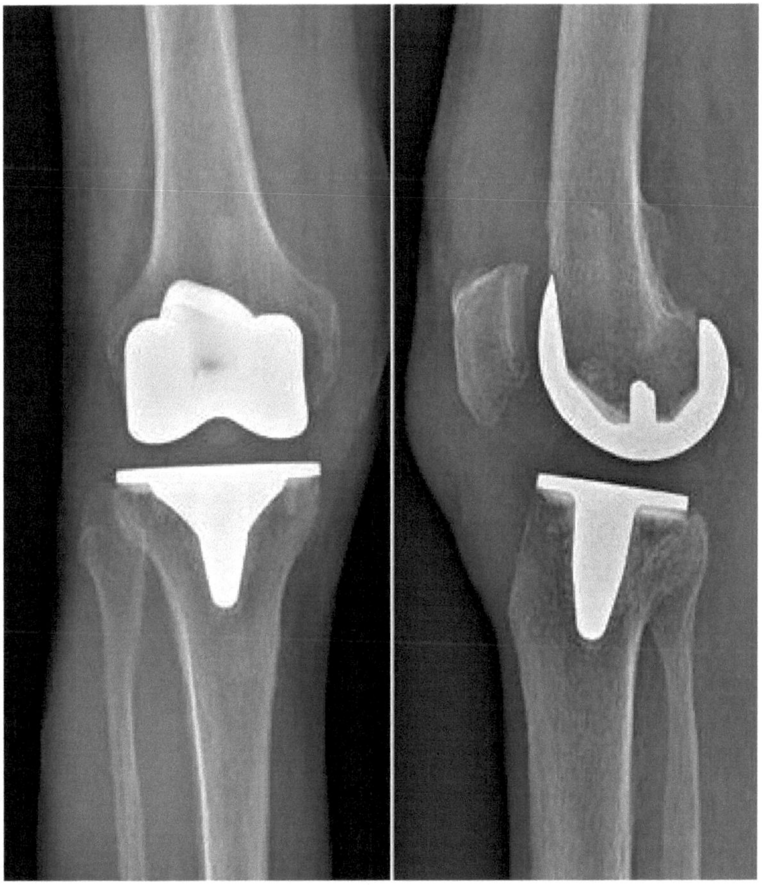

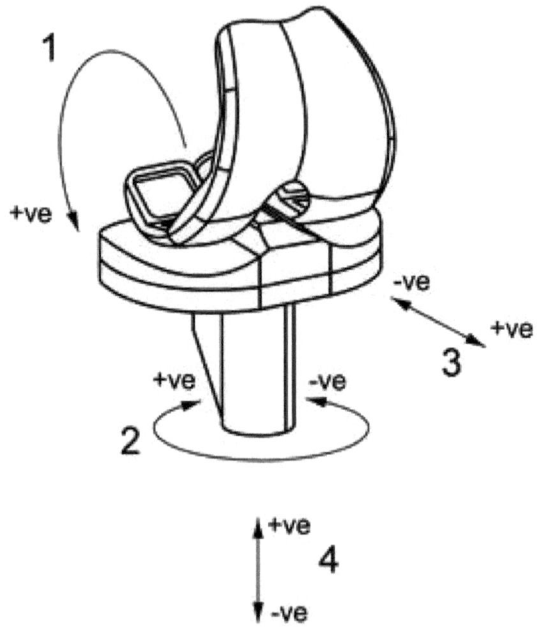
1
+ve
-ve
+ve
3
+ve
-ve
2
+ve
4
-ve

Printed by Books on Demand GmbH, Norderstedt / Germany